STUDY G

FIRE OFFICER'S HANDBOOK OF TACTICS

5TH EDITION

STUDY GUIDE

FIRE OFFICER'S HANDBOOK OF TACTICS

5TH EDITION

JOHN NORMAN

Fire Engineering
BOOKS & VIDEOS

Disclaimer: The recommendations, advice, descriptions, and methods in this book are presented solely for educational purposes. The author and publisher assume no liability whatsoever for any loss or damage that results from the use of any of the material in this book. Use of the material in this book is solely at the risk of the user.

Fire Engineering Books & Videos
110 S. Hartford Ave., Suite 200
Tulsa, Oklahoma 74120 USA

800.752.9764
+1.918.831.9421
info@fireengineeringbooks.com
www.FireEngineeringBooks.com

Senior Vice President: Eric Schlett
Operations Manager: Holly Fournier
Sales Manager: Joshua Neal
Managing Editor: Mark Haugh
Production Manager: Tony Quinn
Developmental Editor: Chris Barton
Book Designer: Susan E. Ormston
Cover Designer: Elizabeth Wollmershauser

ISBN13 978-1-59370-438-4

Printed in the United States of America

2 3 4 5 25 24 23 22 21

For all the serious students of firefighting who wish to be the best at protecting humankind.

And for all the instructors who help us in that endeavor, especially

Dave Ballenger

Donald Burns

Brian Fahey

Brian Hickey

and

Ray Meisenheimer

CONTENTS

INTRODUCTION

THIS study guide is intended to be used in conjunction with the textbook, *Fire Officer's Handbook of Tactics, Fifth Edition*, by John Norman. It is important to realize that this subject is a comprehensive one, and no fire officer will be able to learn everything there is to know about tactics just from reading the text and using this study guide. Department policies and procedures must be followed, as well as any and all codes and standards that are applicable.

However, when studying, this guide will enhance the lessons put forth in the text. It will be a benefit to the instructor in developing tests and to the students in reviewing materials in conjunction with the text.

The study guide has been divided according to the text, with a total of 23 chapters, plus a final exam. You will find questions, answers, and references to the pages in the text where the most pertinent information appears for each question. Questions in the final examination may require the student to synthesize information from more than one chapter in order to reach the correct answer, much as problems on the fireground require a variety of information to resolve.

One final note on study: The more sources of input one has, the more the mind absorbs. Reading the text is only one step and uses only one sensory input—the eyes. Another useful technique is to answer a series of questions about the material after reading it to test your understanding. That is what this study guide does. It has long been recognized that more sources of input help the mind absorb information. I am not talking about putting the text under your pillow and hoping for osmosis. Reading the text aloud while studying can also be helpful. Listen to the audio book while in your car, or read the text into a recording device and play it on your mobile device. As a student who was very successful in taking promotional exams, I used all of the above techniques. I swear by repeatedly listening to audio books. This works the same way as you learn the words to your favorite songs: through immersion. Find the techniques that work best for you, and stick with them. The rewards of truly knowing your trade are great, and if promotion into a leadership position is your goal, you will have a long leg up on your competition.

ABOUT THE AUTHOR

Deputy Assistant Chief (Ret.) John Norman is a veteran of more than 40 years in the fire service, including 27 years with the FDNY, the majority of that time in the Special Operations units. Chief Norman served as a firefighter with Engine Co. 290 and Ladder Co. 103 in Brooklyn; Rescue Co. 3, which covered the Bronx and Harlem; and helped organize Hazardous Materials Company 1 at its inception. As a lieutenant he served with Rescue Co. 2, which covered all of Brooklyn, and as a captain he was the company commander of Rescue Co. 1, which is located in the heart of Manhattan's high-rise district, just off Times Square.

While still assigned to Rescue 1, John worked with the Safety Command and the Fire Commissioner's High Rise Fire Safety Task Force, investigating and then developing high-rise fire safety legislation in the wake of the fatal Vandalia Avenue and Macaulay Culkin fires, which killed three firefighters and six civilians. Promotion to battalion chief in 1999 brought assignment to Battalion 10 and the Y2K Preparation Task Force before assignment to Battalion 16.

In the days after September 11, 2001, John was designated as search and rescue manager for the World Trade Center site. He operated in that position for two months before assuming the job of chief of special operations on a full-time basis. He was promoted to deputy chief in June of 2003, remaining in Special Operations. In August 2004, Chief Norman was promoted to deputy assistant chief, and once again designated as chief of special operations as well as a citywide tour commander. He retired from the FDNY in 2007. During his career he was decorated for valor nine times.

Chief Norman attended Oklahoma State University, majoring in fire protection technology. After college he worked as a fire protection engineer for a major NYC. fire protection contractor, designing sprinkler, standpipe, and other fire protection systems for many buildings in the metro New York area, including such famous addresses as 666 Fifth Avenue and portions of the ill-fated World Trade Center. He earned a bachelor's degree in fire service administration from Empire State College of the State University of New York. He attended the FDNY/USMA West Point Combating Terrorism Leadership Program and is a graduate of the FDNY/Columbia University Fire Officers Management Institute. He has testified as an expert witness concerning fire operations in both state and federal court cases. He served as a consultant to the city of Chicago in the wake of the tragic Cook County Office Building fire.

Chief Norman has been an adjunct instructor at the FDNY Fire Academy on Randall's Island, teaching in the Chief Officers Development, Confined Space, Building Collapse, and Foam Coordinator's/Flammable Liquid Training courses. As a task force leader with the NYC Urban Search and Rescue Task Force, John has taught in the Rescue Specialist Training course given by FEMA to the nation's 28 USAR teams. He is also the author of *Fire Department Special Operations*. He has written for such fire service magazines as *WNYF*, *Fire Engineering*, and *Firehouse*, and served on validation committees for the International Fire Service Training Association manuals on the following subjects: ventilation, rescue, building construction, and terrorism. He lectures nationally on a wide variety of fire and rescue topics. He can be reached at www.chiefnorman.com.

PART I

GENERAL FIREFIGHTING TACTICS

01 GENERAL PRINCIPLES OF FIREFIGHTING

Questions

1. What is the most basic principle of firefighting?

A. Human life should take precedence over all other concerns.

B. Saving property helps save lives.

C. Firefighters should protect themselves at all costs.

D. Property and human life should receive equal attention.

2. As a firefighter who finds the need to choose between rescuing a victim or extinguishing the fire, what factors impact most heavily on your decision?

1. Is sufficient personnel available to extinguish the fire and conduct rescues simultaneously?
2. Can access be made to the victim that does not require hoseline support?
3. Are there other units in quarters that can be called to perform the rescue?
4. Would putting the fire out eliminate the need for rescue?
5. What is the fire loading?

A. 1 and 2 only

B. 1, 2, 3, and 4 only

C. 1, 2, and 4 only

D. all of the above

3. You arrive as the officer in command of the first-due engine at a fire in a four-story apartment house at 3 a.m. You find a fire venting from four ground-floor windows, and heavy smoke showing from all floors. Your crew consists of a driver, two firefighters, and yourself. The next unit to arrive will likely be 6–8 minutes behind you. Numerous bystanders are reporting multiple people trapped in the 24 apartments in the building. What is the most correct action to order?

A. Order the hoseline placed between the fire and the trapped occupants.

B. Order the unit's portable ladder raised to remove the trapped occupants.

C. Order the members to begin search efforts on all floors.

D. Order the members to begin ventilating the structure.

4. You arrive as the officer in command of the first-due ladder company at a serious fire on the second floor of a six-story, ordinary construction apartment house and find people hanging out of numerous windows calling for rescue. List the following selection of victims in the order that they should be removed.

1. persons in the immediate vicinity of the fire
2. persons on the fire floor, remote from fire
3. persons on the floor directly above the fire
4. persons on the top floor
5. persons below the fire
6. persons three floors directly above the fire, which is two floors below the top floor

The correct order is:

A. 1, 4, 3, 6, 2, 5

B. 1, 2, 4, 3, 6, 5

C. 1, 3, 4, 2, 6, 5

D. 4, 1, 3, 2, 6, 5

5. "The lives of the occupants of the immediate fire area must always be the highest priority where personnel is insufficient to accomplish rescue and removal simultaneously." This statement is:

A. Correct. Occupants of the fire area must always be the first rescued.

B. Correct. If the victims are in fact savable.

C. Incorrect. Begin easiest rescues first, then progress to most difficult.

D. Incorrect. Rescues must begin where the greatest numbers of occupants are seen.

6. "Interior structural firefighting requires an aggressive attack at every incident." The above statement is:

A. Correct as written.
B. Correct, except that firefighters should never enter vacant buildings.
C. Incorrect, since vacant buildings should be treated with extreme caution.
D. Incorrect, since aggressiveness should be avoided at all costs.

7. Which of the following describes the proper sequence of actions at most structural fires?

A. size-up, locate, confine, extinguish
B. locate, protect exposures, rescue, size-up
C. confine, locate, extinguish, overhaul
D. confine, rescue, extinguish, protect exposures

8. At 3 o'clock in the morning you arrive simultaneously with the first engine to find fire venting from two windows on the ground floor of a three-story brick and wood joist warehouse. The fire is exposing a windowless fireproof cold storage warehouse across a 20-ft-wide alley. Where should the first hoseline be placed?

A. in the alley between the fire building and the exposure
B. inside the fire building
C. inside the exposed building
D. in front of the fire building, supplying a master stream

9. Early one Sunday morning you arrive to find heavy fire blowing out of all the front windows of a two-story private house. Two cars are in the driveway, and neighbors tell you that a family of five live in the home, but have not been seen since the previous evening. What is the most correct action to take?

A. Begin an aggressive interior search.
B. Begin defensive operations with emphasis on protecting exposures, since there is no possibility of survivors.
C. Begin an aggressive interior attack, knocking down as much fire as rapidly as possible, then searching for survivors.
D. Begin an aggressive interior attack, knocking down as much fire as rapidly as possible while simultaneously searching for survivors.

10. **You arrive as the officer in command of the first-due engine company to find fire venting from one window on the ground floor rear of a two-story private house. Your crew consists of a driver, two firefighters, and yourself. You find a teenage girl calling for help from the front window on the second floor. What is the most appropriate action to order?**

    A. Commit all personnel to the rescue effort utilizing the pumper's ground ladder.

    B. Commit all personnel to the rescue effort using an inside approach.

    C. Commit all personnel to fire control with a hoseline, then search.

    D. Commit two members to stretch a hoseline to protect the interior stair, while the driver and one firefighter place the pumper's ground ladder to remove the teenager.

02 SIZE-UP

Questions

1. What is size-up?

A. Size-up determines the size of the fire.
B. Size-up is where you determine how many victims are in a structure.
C. Size-up is a continuing evaluation of all the problems and conditions that affect the outcome of the fire.
D. all of the above

2. Who performs size-up?

A. The officer of each engine company should perform size-up.
B. The first firefighter to enter a structure should perform the size-up.
C. Size-up should be performed by all members, to varying degrees.
D. Only the officer in charge should conduct the size-up.

3. When should size-up be performed?

A. Size-up should be conducted as soon as the alarm is received.
B. Size-up should begin upon arrival of the first officer at the scene.
C. Size-up can only be conducted after entering a structure.
D. Size-up should begin upon the receipt of the alarm and continue until the incident has been controlled.

4. **Given the following alert message, what element of the size-up can you ascertain? "Attention, Engines 1, 2, and 3; Ladder 1; Chief 1: Respond to a house fire at 158½ Main Street. Fire reported on the third floor. The time is 1430 hours."**

   A. Three engines, one ladder, and one chief have been assigned.

   B. There is no life hazard present.

   C. The fire is in a frame home.

   D. Fire is on the third floor.

   E. all of the above

5. **In *your community*, on which side of the street would the even-numbered homes be found? Select only one choice.**

   A. on the east side on a north–south street

   B. on the west side on a north–south street

   C. on the north side on an east–west street

   D. on the south side on an east–west street

6. **A rule of thumb for collapse anticipation in ordinary construction or wood-frame buildings is that if fire has heavily involved an area for ____ minutes, fire forces should be withdrawn.**

   A. 5

   B. 10

   C. 15

   D. 20

7. What are the main elements of the traditional 13-point or COAL WAS WEALTH size-up?

1.	**C**:	company	construction	core construction
2.	**O**:	occupancy	ordinary	officer
3.	**A**:	allowed personnel	allied personnel	apparatus and personnel
4.	**L**:	life hazard	location	line placement
5.	**W**:	windows	water supply	wall
6.	**A**:	airflow	attack	auxiliary appliances
7.	**S**:	stairs	street conditions	supply lines
8.	**W**:	weather	warnings	work area
9.	**E**:	exhaust opening	exposures	extreme conditions
10.	**A**:	air cylinders	allocation	area
11.	**L**:	loss	ladder selection	location and extent of fire
12.	**T**:	trusses	time	threat
13.	**H**:	heat	hydrant location	height

8. A modernization of the traditional 13-point size-up would combine ___________ and _______________, and add ________________________ to the listing.

A. area, height, wind speed
B. area, height, hydrant locations
C. area, height, hazardous materials
D. area, height, water volume

9. What is the best method of reducing the life hazard before the fire?

A. improving exit facilities
B. specifying fire door installation
C. reducing occupancy load
D. installing a wet pipe sprinkler system

10. **Knowing the type of occupancy involved can tell an officer much about a given incident. Name five key variables strongly tied to occupancy.**

    1. potential life hazard
    2. wooden shingles
    3. presence of large, open floor spaces or small rooms
    4. presence of hazardous materials
    5. exposure hazards
    6. degree of fire loading
    7. possible presence of truss construction (large, open floor areas)

    A. 1, 2, 4, 6, 7
    B. 1, 3, 4, 5, 6
    C. 1, 3, 4, 6, 7
    D. 1, 4, 5, 6, 7

11. **Statistics have shown that the greatest civilian life hazard occurs in which occupancy?**

    A. vacant buildings
    B. residential
    C. stores/offices
    D. public assembly

12. **As far as construction affects the size-up, a fire involving a newer metal deck roof generally _____________ the fire load.**

    A. eliminates
    B. reduces
    C. increases
    D. adds little or nothing to

13. **Which class of construction is most resistant to collapse caused by fire?**

    A. older type Class 1
    B. newer type Class 1
    C. newer type Class 4
    D. older type Class 4

14. **What factors make the 20-minute rule of thumb for predicting collapse less than scientific?**

    A. unknown length of burn
    B. potential use of accelerants
    C. possibility of the presence of lightweight construction
    D. all of the above

15. **Three unusual circumstances may confuse your size-up of the height and area of the building. What are they?**

    A. buildings with cocklofts, buildings with fire walls, and interconnected buildings
    B. building built midblock, buildings with attics, and interconnected buildings
    C. buildings built on grade, wrap-around buildings, and interconnected buildings
    D. buildings built on grade, wrap-around buildings, and buildings with fire walls

16. **Which of the following choices does not represent a special firefighting problems related to location of the fire.**

    A. a cellar fire in a fireproof building
    B. a top-floor fire in a frame apartment complex
    C a fire on the second floor of a three-story windowless warehouse
    D. a fire on the second floor of a 12-story high-rise hotel

17. **What three factors concerning smoke should be prime elements of a size-up?**

    A. location, color, movement
    B. location, smell, taste
    C. smell, color, taste
    D. color, movement, heat

18. A firefighter who needed to report fire extending into the attached row house marked with an X should report that she is in which building?

A. exposure 2
B. exposure 2A
C. exposure 2B
D. exposure D2

19. Which of the following most correctly lists the minimum required fire flow for each occupancy below, expressed in gpm per 100 sq ft, in a modern, nonvented fire situation?

A. private home: 10, stationery store: 15, mattress factory: 20 (gpm/100 sq ft)
B. private home: 10, stationery store: 20, mattress factory: 30 (gpm/100 sq ft)
C. private home: 15, stationery store: 20, mattress factory: 40 (gpm/ 100 sq ft)
D. private home: 15, stationery store: 35, mattress factory: 60 (gpm/ 100 sq ft)

20. Which of the following is not a type of auxiliary appliance?

A. dry chemical spray-booth system
B. dry chemical fire extinguisher
C. bulk oil storage foam system
D. standpipe

21. Complete the following descriptions of the effects that weather can have on firefighting.

1. High heat and humidity rapidly ______________________________.
2. Below freezing temperatures result in ______________________ and causes ______________________________.
3. High winds can _________________________ or force fire__________________________.

22. **In the event a department encounters a sewer trench in the fire block between the roadway and the fire building, the ___________ should enter the block while the __________ remains out at the intersecting street.**

    A. aerial device, pumper
    B. pumper, aerial device

23. **On a hot, humid day, you are the incident commander at a fire in a two-story Class 5 frame house with a cellar. Your first attack hoseline has entered the structure and has knocked down a lot of fire. The crew returns to the street with their air cylinders depleted, and the officer approaches to inform you that the main body of the fire has been knocked down, but there is still debris smoking that needs overhauling. You consult your dispatcher and find that you are 20 minutes into the alarm. What is the most correct order to issue?**

    A. Have all units evacuate the building.
    B. Have only the original unit evacuate the building.
    C. Have the original unit replace their cylinders and complete the overhaul.
    D. Have a fresh unit enter the structure to complete the overhaul.

24. **All of the following are advantages of the time mark system except for which choice?**

    A. It avoids tunnel vision.
    B. It refocuses the IC's attention on the elapsed time of burn.
    C. It forces the IC to evaluate what progress is being made.
    D. It determines whether a life hazard is present.

25. **Which of the following factors is responsible for the destruction of more buildings than any other construction impacts?**

    A. the degree of compartmentation
    B. the degree the building contributes to the fire load
    C. the presence of hidden voids
    D. the ability to resist collapse

03 ENGINE COMPANY OPERATIONS

Questions

1. Hose streams may be used for a variety of purposes, including which of the following?

1. fire extinguishment
2. exposure protection
3. controlled burning
4. forced ventilation of gases
5. absorbing toxic fumes

A. 1, 2, and 4 only
B. 1, 2, 3, and 4 only
C. 1 and 2 only
D. all of the above

2. Which of the following is not one of the common methods of structural fire attack?

A. direct attack
B. indirect attack
C. combination attack
D. exposure protection

3. Which of the following is a potential problem encountered during incipient stage operations?

A. sudden ignition of an aerosol spray can
B. sudden fireball from a pooled liquid struck with a fog stream
C. fire exceeding the capacity of a midsized handline to control
D. backdraft explosion

4. Which choice does not properly describe an activity of the nozzle team?

A. Attempt to locate and account for occupants and gather information from them.
B. Survey the layout of the structure for alternate escape routes.
C. Stretch dry hose to a safe location.
D. Arrive at the fire area with sufficient hose to cover the first room.

5. Among the reasons for selecting the combination method of attack on a fire, which reason is of the greatest importance?

A. It avoids steam production around the nozzle team.
B. It disrupts the thermal balance tremendously.
C. It puts firefighters in the best position to save lives.
D. It exposes firefighters to the least danger.

6. Prior to beginning the actual attack on a free-burning fire, which actions should the nozzle team take?

1. Attempt to locate and account for occupants.
2. Survey the structure for alternate escape routes and other fires in remote areas.
3. If you are approaching from below, take a quick look at the floor below the fire to get the layout.
4. Begin ventilation above the fire.

A. 1 and 2 only
B. 1, 2, and 3 only
C. 2, 3, and 4 only
D. all of the above

7. At a working cellar fire in a two-and-a-half-story frame house, the first line should usually stretch dry hose to which area before charging the hoseline?

A. the front door
B. the top of the unenclosed cellar stair
C. the top of the enclosed cellar stair
D. the bottom of the enclosed cellar stair

8. The method of attack best suited for the free-burning stage is:

A. direct attack
B. indirect attack
C. combination attack
D. alternative attack

9. Rollover is an indicator of impending:

A. flashover
B. backdraft
C. smoke explosion
D. all of the above

10. Finding a heavy fire on the fifth floor of a six-story ordinary construction apartment house, the IC ordered the members of the second-arriving engine to assist the first engine in stretching their hoseline, even though it was clear that a second line would be required. The officer's actions were:

A. Correct. Get the first line in operation quickly, before stretching additional lines.
B. Incorrect. The second line is needed and should be stretched immediately.
C. Incorrect. The members of the second engine should immediately begin evacuating the occupants.
D. Correct. Two companies are needed to operate a hoseline under heavy fire conditions.

11. Three strategic options in the firefighting plan include all but which choice?

A. Attack the exposures.
B. Make an offensive attack.
C. Establish defensive positions.
D. Take no action at all.

12. What factors are primarily responsible for most fire spread to exposures?

1. radiation
2. conduction
3. convection
4. direct flame contact

A. 1 only
B. 1 and 2 only
C. 1 and 4 only
D. all of the above

13. An officer in command of the first attack line finds that the line is unable to advance into the fire area due to heavy fire and tremendous heat. The officer should promptly request all but which of the following actions?

A. Call for additional ventilation opposite the line.
B. Call for an increase in pump pressure on the first line.
C. Call for a backup line.
D. Call for an emergency evacuation of the building.

14. The hoseline referred to in question 13 is still unable to advance after trying all of the correct options cited above. What actions should you not consider next?

A. Change the direction of the attack.
B. Use handlines to darken heavy fire.
C. Breach a wall to darken fire.
D. Use distributors, cellar pipes, high-expansion foam, etc.

15. Officers in command of operations expect to see some visual indication that hoseline advance is successful. All of the following items should indicate that a line is successfully hitting fire EXCEPT:

1. flames darkening down
2. change in smoke color
3. change in smoke volume
4. change in smoke movement
5. steam production
6. cessation of collapse of a wood-frame home

A. none of the above
B. 5 and 6 only
C. 3 and 6 only
D. 6 only

16. An elevating platform's master stream can cover the lower three floors of a structure for a length of how many feet (frontage) if properly positioned?

A. three times the length of the stream
B. 75 ft
C. 100 ft
D. 150 ft

04 HOSELINE SELECTION, STRETCHING, AND PLACEMENT

Questions

1. **Four main factors affect the selection and placement of a hoseline. Which is incorrectly stated?**

   A. occupancy
   B. street frontage
   C. construction
   D. location and extent of the fire

2. **"With the new 1¾- or 2-in. hose and automatic nozzles, we can maintain the same flow as with a 2½-in. line." This statement is:**

   A. correct as written
   B. correct, if the nozzle pressures are equal
   C. incorrect, if the nozzle pressures are equal
   D. incorrect, if a proper size nozzle is used on the 2½-in. line

3. **Compared with a smaller line, a larger hoseline exhibits all but one of the following characteristics. Which one is not an attribute of a larger diameter line?**

   A. has greater flow potential
   B. has greater flexibility
   C. has greater reach and stream impact
   D. puts out fire and cools well ahead of members

4. Residential occupancies have several characteristics that affect hoseline selection. Which of the following is not one of them?

A. time of day
B. need for speed
C. low fire loading
D. presence of dividing walls or partitions

5. All of the following areas present similar degrees of compartmentation and light fire loading except which choice?

A. a high school classroom
B. a junior high storage area
C. a fireproof hotel
D. a hospital patient care area

6. What factors determine the required diameter of the attack line?

1. occupancy
2. area of the building (potential fire area)
3. size of the fire
4. time of day

A. 3 only
B. 1 and 2 only
C. 1, 2, and 3 only
D. all of the above

7. What factors determine the required length of the attack line?

1. setback from the street
2. height and area of the building
3. presence or absence of an open stairwell
4. presence or absence of a standpipe system
5. occupancy

A. all of the above
B. 1, 2, 3, and 4 only

C. 1, 2, and 3 only

D. 1 and 2 only

8. How much hose should be stretched for a fire on the top floor of a three-story factory building, 200 ft long × 200 ft wide (*not* tall), with the hydrant 100 ft from the front entrance?

A. 500 ft of 1¾-in. hose

B. 500 ft of 2½-in. hose

C. 600 ft of 2½-in. hose

D. 900 ft of 2½-in. hose

9. Firefighters stretching to a fire on the fifth floor via a wide, open stairwell in an old apartment house should ensure that at least how many lengths of hose are present inside the building? (Assume the stairway to be right at the front door.)

A. two

B. four

C. five

D. six

10. Stretching hose via a rope on the outside of the building has all of the following advantages over stretching around a staircase except for which of the following?

A. It is faster.

B. It allows the members on the first line to attack from the fire escape with the wind at their backs.

C. It requires fewer members to accomplish.

D. It provides a ready means to stretch additional lines.

11. The degree of danger to a brick or concrete building from an exposure fire depends on:

1. the number and size of windows
2. the fire resistance rating of the walls
3. the proximity to the fire building
4. the response of an adequate number of fire department ladder companies

A. 1 and 3 only

B. 2 and 4 only

C. 1, 2, and 4 only

D. all of the above

12. What provides the greatest protection against radiant heat?

A. water curtains
B. distance
C. combustible construction
D. light-colored or reflective paints

13. Which is not one of the main functions of a nozzle?

A. to control the flow of water
B. to increase the velocity of the water
C. to divide the stream into fine droplets
D. to give the stream its shape

14. For maximum effectiveness, the discharge opening of a nozzle should not exceed what percentage of the supply diameter?

A. 25%
B. 50%
C. 75%
D. 100%

15. The bent applicator pipe in useful for discharging all but which agent?

A. dry chemical
B. water
C. aqueous film-forming foam
D. fluoroprotein foam

16. Most fires in residential occupancies are best handled with a midsized handline. Which statement below does not properly state a reason why this is true?

A. Residences are generally more confined than stores.
B. Residences have relatively light fire loads.
C. Residences are all single-story fires.
D. Residential life hazards demand speed of water application and a rapid advance.

17. Most fires in commercial occupancies should be fought with 2½-in. hose. Which statement below properly states a reason why this is true?

A. The fire loading is heavier in commercial buildings than in factories.
B. The size of the fire in a commercial building is more readily determined than in a house fire.
C. The floor areas demand a longer-reaching, harder-hitting stream.
D. Often there are less flammable materials around that can accelerate a minor fire.

18. An advantage of 2½-in. hose over smaller diameter lines is correctly stated in which choice?

A. A larger flowing line has more reach than a smaller line at the same pressure.
B. A larger flowing line has greater penetration than a smaller line at any pressure.
C. A larger line takes the same number of personnel to hold as a smaller line.
D. A larger line takes four people to hold.

19. A fully involved commercial occupancy 20 ft wide × 50 ft deep, common to many smaller strip malls, would require approximately how much flow to ensure rapid knockdown and extinguishment?

A. 100 gpm
B. 200 gpm
C. 350 gpm
D. 500 gpm

20. Stretching a hoseline to the correct location is easier said than done. Which of the following statements regarding hoseline stretching would be most correct?

A. Be sure an uncharged hoseline does not run under any doors. Use a rug or a welcome mat to chock doors closed.
B. After the nozzle operator has removed the working length, he should step off two or three paces and let the backup person remove the remainder of the hose.
C. The backup person should bring as much spare hose as possible into the fire building where it will be ready to advance.
D. If the nozzle team encounters a narrow stairwell, they should stretch the hose up the well to minimize the number of lengths required to reach upper floors.

21. A senior firefighter in an engine company relayed her years of experience to a newer member. She made the following statements regarding hose stretching to upper floors. She was correct in all but which choice?

A. If a wide stairwell is encountered, it is best to stretch the line in the stairwell. The presence of such a well must be relayed to those stretching the hose to avoid stretching too much hose.

B. When stretching up a stairwell, do not stretch more than two lines in the well to avoid them getting tangled.

C. When stretching to upper floors via a wide well, one length of hose will usually reach up to the fifth floor. The length must be supported just above the coupling with a rope or hose strap.

D. If an elevator is installed in the well, it is best to use a rope to haul the line up the outside of the building if the fire is above the third floor.

22. Evaluate the following statements made about alternate approaches for stretching hoselines and select the most correct.

A. When stretching lines up the exterior, several methods are possible. All should focus on bringing the line into the building directly on the fire floor for speed.

B. If a line is to be stretched up a fire escape, it must be carefully coordinated with inside members to avoid opposing hose streams.

C. Stretching a line up an aerial or platform can speed operations, and is especially useful for attack if there is a gated outlet on the platform or if a line can be connected directly to a ladder pipe.

D. If a rope is used to haul a line up the outside, avoid stretching more than one line this way so they will not become entangled.

05 WATER SUPPLY

Questions

1. **How would a gauge inserted at point A read, compared to a similar gauge at point B?**

   A. the same

   B. lower

   C. higher

   D. the answer is indeterminable

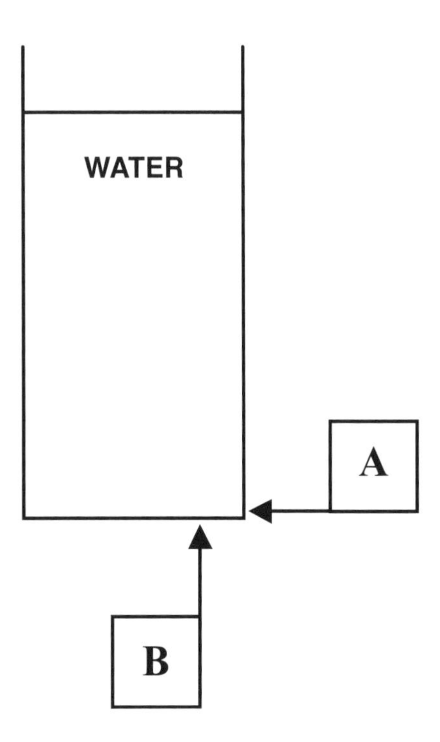

2. If a pumper were discharging into the hoseline pictured below at a pump pressure of 150 psi, what would the nozzle pressure at gauge D read?

   A. 75 psi
   B. 100 psi
   C. 125 psi
   D. 150 psi

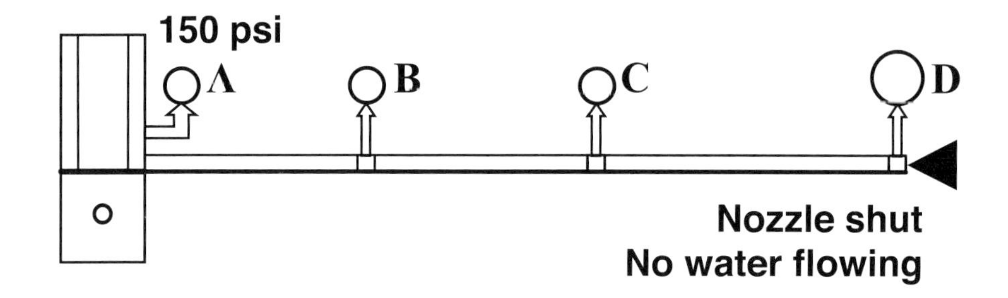

3. What should the gauge at point X be reading?

   A. 2 psi
   B. 8 psi
   C. 12 psi
   D. 18.4 psi

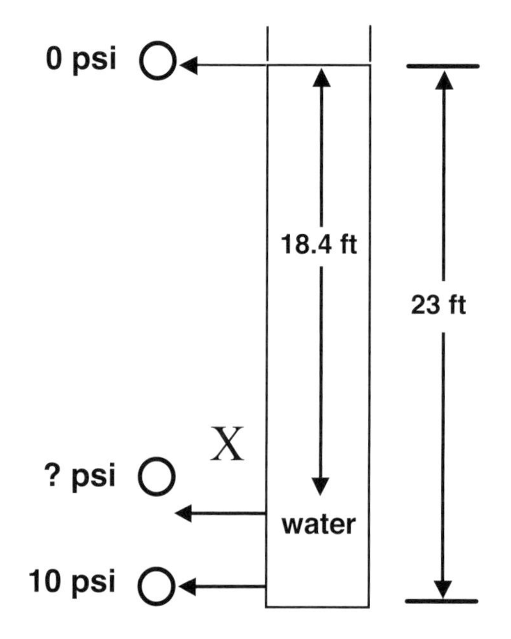

4. Gauges A–D all read the same pressures. They illustrate what type of pressure:

A. elevation
B. gravity
C. head
D. all of the above

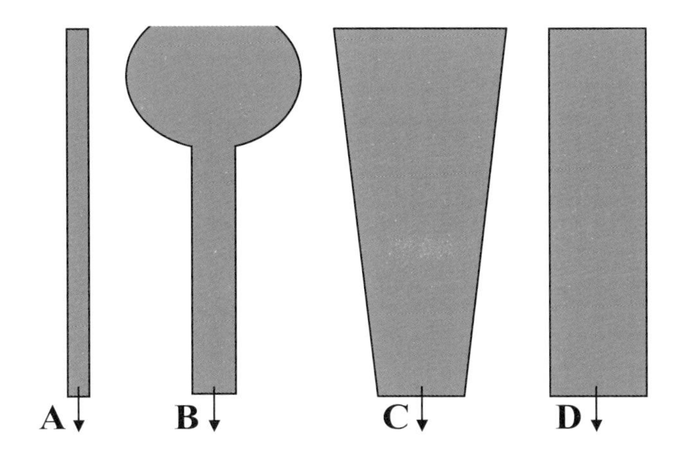

5. With water flowing at hydrant A, what should the residual pressure read at hydrant B?

A. 90 psi
B. 95 psi
C. 170 psi
D. 175 psi

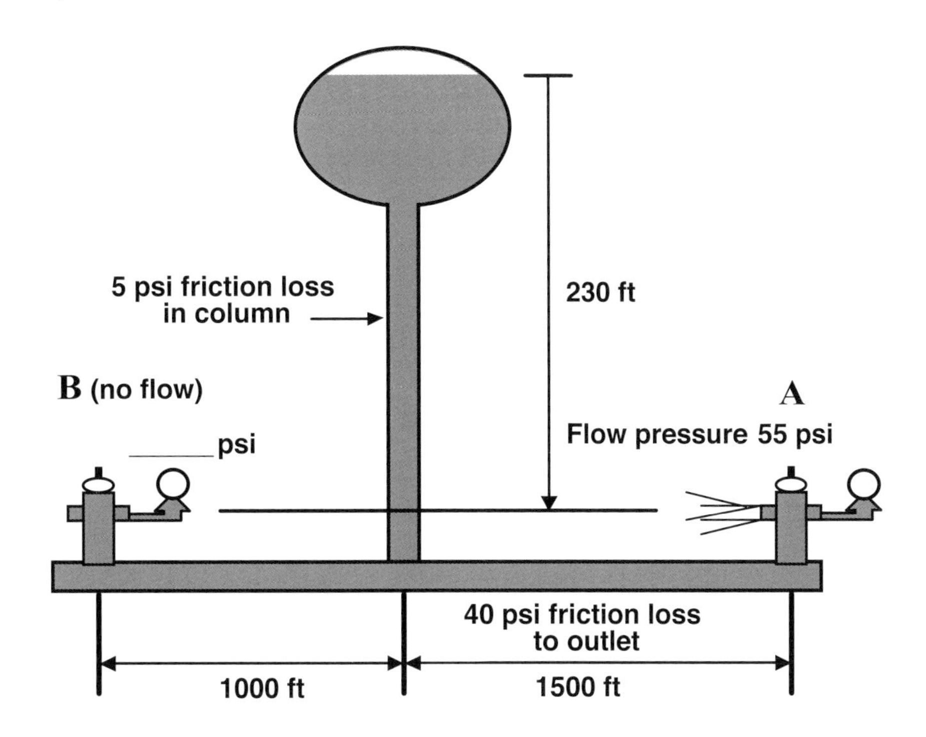

6. At high flows, most of what is commonly called friction loss is due to which of the following?

A. friction between the water and the hose lining
B. turbulence at the hose couplings
C. turbulence within the water flow
D. gently curving bends in the hose

7. Which of the following factors does not affect the water supply capability of a given layout?

A. the capacity of the pump
B. the capacity of the hose
C. the capacity of the water source
D. the available staffing

8. The minimum incoming pressure that a fire pump should receive is:

A. 5 psi
B. 10 psi
C. 25 psi
D. 150 psi

9 Where long distances or large flows are required of a relay operation, which choice doesn't solve the water delivery problem?

A. changing the size of the discharge nozzle
B. laying multiple supply lines
C. using large diameter supply lines
D. using a larger capacity pumper at the source

10. What are some of the advantages of a telescoping platform over an aerial ladder for elevated master stream use?

1. greater tip-loading capacity
2. greater discharge capacity
3. greater range of motion of the stream
4. higher elevation
5. the ability to reach over obstructions, such as wires, in front of a fire building

A. 1 and 2 only
B. 1, 2, and 3 only
C. 1, 2, 3, and 4 only
D. all of the above

11. By dividing the flow of water from one line into two similarly sized lines, the friction loss is reduced to what percentage range of the loss for a single line?

A. 25–35%
B. 45–55%
C. 60–70%
D. 75–85%

12. One cubic foot of water weighs most nearly:

A. 8.5 lb
B. 52 lb
C. 62 lb
D. 72 lb

13. Elevation pressure loss when pumping to an upper floor can be estimated at how many psi per floor, given 10 ft per story?

A. 0.434 psi
B. 2 psi
C. 5 psi
D. 10 psi

14. The maximum height from which it is theoretically possible to draft water using a centrifugal fire pump is approximately:

    A. 14.7 ft
    B. 20 ft
    C. 25 ft
    D. 33 ft

15. What effect would tripling the length of a hoseline have on the total friction loss if the flow remained the same?

    A. none
    B. one-third the loss
    C. triple the loss
    D. nine times the loss

16. What effect does increasing the diameter of a hoseline have on the flow if the friction loss per foot remained constant?

    A. none
    B. doubled flow
    C. increased flow
    D. reduced flow

17. Of the following types of water pressure, which is of most use to firefighters considering a hydrant-to-fire hose stretch?

    A. static
    B. flowing
    C. residual
    D. head

18. Which of the following hydraulic principles is incorrectly stated?

    A. Regardless of the diameter of a hose, the more water flowing through it, the higher the friction loss will be.
    B. Regardless of the length of a hoseline, at a given flow, the friction loss remains the same.

C. Regardless of the length of a hoseline, if there is no flow, the pressure remains the same throughout the line.

D. Regardless of the flow, by increasing hose diameter, the same amount of water can be moved farther with the same amount of energy.

19. Applying hydraulic knowledge to water supply problems allows firefighters to correctly supply hose streams. Which statement below is least correct regarding hydraulics?

A. By increasing the diameter of a hoseline, more water can be delivered at the same pump discharge pressure.

B. By increasing the length of a hoseline, more water can be delivered at the same pump discharge pressure.

C. By increasing the discharge rating of a pumper, more water can be delivered at the same pump discharge pressure.

D. By increasing the discharge rating of a pumper, more water can be delivered at a higher pump discharge pressure.

06 SPRINKLER SYSTEMS AND STANDPIPE OPERATIONS

Questions

1. According to NFPA statistics, sprinklers have successfully controlled or extinguished approximately what percentage of the fires at which they are present?

    A. 65%
    B. 75%
    C. 85%
    D. 95%

2. In more than 90% of all sprinkler operations, the fire is successfully contained by how many sprinklers?

    A. 1 or 2
    B. 5 or 6
    C. 10–12
    D. 20–25

3. How many cases of multiple deaths due to fire have occurred in buildings where operational wet pipe sprinklers were present throughout? (Exclude the World Trade Center attack of 2001.)

    A. thousands
    B. hundreds
    C. dozens
    D. none

4. Which of the following are correct reasons for allowing the sprinklers to continue to operate until the fire has been extinguished?

   1. The sprinklers and hose streams combined may be holding a fire in check.
   2. There may be a second remote fire in the building that firefighters are unaware of.
   3. The water damage from sprinklers is helpful to salvage efforts.
   4. In the case of explosions, there is often broken sprinkler piping that is putting water directly on the seat of the fire.

   A. 1 only
   B. 1 and 2 only
   C. 1, 2, and 3 only
   D. all of the above

5. A typical sprinkler head operating at 50 psi will flow approximately how many gpm?

   A. 12
   B. 20
   C. 30
   D. 40

6. The same sprinkler operating at 5 psi will flow approximately how many gpm?

   A. 10
   B. 17
   C. 22
   D. 27

7. What will be the effect of a pumper drawing water from the same source as the sprinkler main?

   A. no effect
   B. increased flow to sprinklers
   C. decreased flow to sprinklers
   D. increased flowing pressure to sprinklers

8. What is the best way to prevent a drop in sprinkler flow?

A. Stretch a supply line to the fire department's Siamese connection.
B. Stretch a handline to the seat of the fire.
C. Supply an elevating platform's master stream.
D. Supply the system with multiple lines of the largest size hose possible.

9. What is the minimum supply that should be stretched to a Siamese?

A. one 2½-in. line
B. two 2½-in. lines
C. one 3-in. line
D. one 5-in. line

10. Which unit should supply the sprinkler Siamese at a large working fire in a sprinklered building?

A. the first-arriving engine
B. the second-arriving engine
C. a multiple-alarm or mutual aid engine
D. an engine other than one supplying hose streams

11. What pressure should the pumper supply to lines feeding most sprinkler Siameses?

A. 100 psi
B. 150 psi
C. 200 psi
D. 250 psi

12. What pressure are most new sprinkler systems normally tested to withstand?

A. 100 psi
B. 150 psi
C. 200 psi
D. 250 psi

13. All but one of the following are correct reasons why fires in sprinklered buildings seem more smoky than fires in unsprinklered buildings. Which choice lists only correct answers?

1. Unsprinklered fires are more often free-burning fires.
2. Sprinklers do not allow complete combustion, creating carbon monoxide (CO).
3. Sprinklers cool the fire gases, making them less buoyant.
4. Sprinkler spray patterns push gases down like a fog nozzle.
5. Fires in unsprinklered buildings burn with less intensity.

A. 1, 2, 3, and 5 only
B. 1, 2, 3, and 4 only
C. 1, 2, 4, and 5 only
D. all of the above

14. Which of the following isn't true of sprinklered fires?

A. no need for increased visibility
B. increased carbon monoxide production
C. sinking fire gases
D. smoke being pushed down

15. What unusual circumstance involving a sprinkler discharge is likely to cause a lot of grief for fire department personnel?

A. heavy smoke impeding firefighters' advance
B. low water pressure drawing water away from firefighters' hoselines
C. sprinklers that discharge after firefighters have begun operating
D. sprinklers that discharge before firefighters arrive

16. Where is this unusual circumstance likely to occur?

A. in buildings with complete wet pipe sprinkler systems
B. in buildings with complete dry pipe sprinkler systems
C. in buildings with partial sprinkler systems
D. in buildings with partial standpipe systems

17. What effect do sprinklers located in skylights have on ventilation efforts?

A. Sprinklers improve ventilation by cooling fire gases.

B. Sprinklers improve ventilation by directing fire gases down through existing openings.

C. Sprinklers impede ventilation by heating fire gases.

D. Sprinklers impede ventilation by cooling fire gases and pushing fire gases down.

18. A system of overhead piping filled with water, connected to a water supply, ready to discharge water at once through a heat activated nozzle, is a description of which type of sprinkler system?

A. deluge

B. automatic wet pipe

C. automatic dry pipe

D. pre-action

19. Where is the main drain valve located on a wet pipe sprinkler system?

A. below the check valve

B. above the check valve

20. Where are automatic dry pipe sprinklers usually found?

A. residential buildings

B. strip malls

C. industrial buildings

D. unheated buildings in cold climates

21. What are the primary differences between a dry pipe valve and a wet pipe alarm valve?

1. The dry valve clapper is larger than a wet pipe valve.
2. The dry valve clapper locks open when tripped.
3. Dry valves must be reset manually.
4. Dry valves contain only a small amount of priming water, plus compressed air.

A. all of the above

B. 1, 2, and 3 only

C. 2 and 3 only

D. 2 only

22. What is a likely cause of a false alarm at a dry pipe sprinkler system?

    A. loss of water pressure
    B. surge of air pressure
    C. water columning
    D. loss of air pressure

23. Both deluge systems and pre-action systems are equipped with all but which of the following components?

    A. an OS&Y valve
    B. fire department connection
    C. a valve actuated by detectors
    D. open sprinkler heads

24. A pre-action sprinkler system would most likely be encountered in which occupancy?

    A. computer rooms
    B. aircraft hanger
    C. the freezer of an ice cream factory
    D. a large nursing home

25. What is the best method of determining if a building is protected by a sprinkler system?

    A. prefire inspection
    B. the presence of a Siamese connection
    C. the sound of an alarm bell
    D. water discharge from drain pipes

26. According to NFPA 14, standpipe systems may be of all of the types listed below except one. Which choice does not belong?

    A. automatic wet standpipe
    B. automatic dry standpipe
    C. semiautomatic dry standpipe
    D. pre-action standpipe

27. All but one of the following sources of supply to sprinkler and standpipe systems are acceptable. Which choice does not belong?

A. connection to water mains (with or without pumps)

B. suction tank, with or without pumps

C. gravity tanks

D. pressure tanks

28. At a serious fire on the upper floors of a standpipe-equipped building, all of the following actions should be taken to supply the standpipe system, except which choice?

A. Ensure that any building fire pumps are operating at the proper pressure.

B. Stretch lines into each Siamese connection.

C. Charge the line only if the fire pump cannot maintain adequate pressure.

D. Pump into the standpipe at the first or second floor if the Siamese is defective.

29. A standpipe system designed to provide first aid hose streams for occupant use, as well as to supply a 2½-in. fire stream, is a ____________ system.

A. Class I

B. Class II

C. Class III

D. Class IV

30. Standpipes that provide hose for occupant use must protect the untrained user from excessive pressure. The pressure reducers limit maximum operating pressure in 1¾-in. hose to how many psi in pre-1993 buildings?

A. 50

B. 65

C. 80

D. 100

31. All of the following items are necessary for conducting operations from a standpipe system, with what exception?

A. at least three lengths of 2½-in. hose with a solid tip nozzle

B. hose thread adapters, spanner, and 24-in. pipe wrench

C. door chocks, door latch straps, handlight

D. forcible-entry tools, portable radio

32. A fire on what floor or below should always be approached via walking up the stairway, while floors above this level might be reached via the elevator?

A. fifth floor

B. sixth floor

C. seventh floor

D. eighth floor

33. What should be the primary source of water supply to a standpipe during fire department operations?

A. building fire pumps

B. gravity tank

C. pressure tank

D. fire department pumper via Siamese

07 LADDER COMPANY OPERATIONS

Questions

1. **What are the benefits of having an established operational plan?**

    1. It formalizes thinking, and makes others think in advance of what must be done.
    2. It assigns a degree of priority to each element.
    3. It lets all members, not just the IC, know the plan.
    4. It establishes accountability for one's actions.

    A. 1, 2, and 3 only
    B. 1 and 3 only
    C. 1, 2, and 4 only
    D. all of the above

2. **What are the three specific items that should be provided as part of any operational plan?**

    A. tools, techniques, and training
    B. number of people, plans, and personnel
    C. number of people, tools, and scope of duties
    D. number of people, tools, and training

3. All but which of the following tools should be provided at every work site at a structural fire?

A. a hook or pole of some type (pike or Halligan)
B. a forcible-entry bar of some type (Halligan tool)
C. flat-head axe
D. pick-head axe

4. When developing an operational plan, all of the following elements are important considerations except for which choice?

A. At least six members must be assigned to truck duties immediately at all alarms.
B. For dwelling fires, search is a key responsibility.
C. All members should use the buddy system.
D. The plan must be realistic.

5. Your personnel level dictates that one member of your ladder company crew may have to operate alone for part of an operation. You must consider all of the following factors except which choice when weighing this responsibility?

A. the experience level of the member
B. the ability to act independently and protect himself
C. the availability of others in remote positions who will be able to assist him as needed
D. the presence of a portable radio for this member

6. What is the predominant fire problem nationally?

A. high-rise office buildings
B. strip mall (taxpayers)
C. apartment houses
D. private homes

7. All but one of the following are duties of an interior search team at a private dwelling. Which choice does not belong?

A. forcible entry
B. roof ventilation
C. search of fire area
D. exposing hidden fire

8. **What does the acronym VES stand for?**

    A. vent, expose, search
    B. vent, examine, salvage
    C. vent, enter, salvage
    D. vent, enter, search

9. **At 10 a.m. on a weekday morning, which room of a home is most likely to contain a victim?**

    A. bedroom
    B. living room
    C. kitchen
    D. den/playroom

10. **Which factor below does not affect portable ladder selection and placement?**

    A. reach
    B. width
    C. stored length
    D. material of construction

11. **What is a major advantage of an extension ladder over a straight ladder?**

    A. lighter weight
    B. greater reach
    C. greater strength
    D. easier to reach the correct working length

12. **Where should the tip of a portable ladder be placed if you are planning to enter a window?**

    A. alongside the top of the window on the upwind side
    B. alongside the top of the window on the downwind side
    C. one or two rungs above the windowsill in the window
    D. just at or below the window sill

13. A disoriented member moves into a room to seek an escape route after being cut off by fire. How could members who placed ladders outside the building have best indicated which window to move toward?

A. by venting the top half of the laddered window with the ladder

B. by completely clearing the laddered window of glass and sash

C. by completely clearing the laddered window and leaving the tip two rungs into the window

D. by venting the top half of the laddered window and leaving the tip two rungs into the window

14. NFPA 1932 specifies a heat-indicating label be applied to new ladders, indicating if that area of the ladder has been exposed to how many degrees Fahrenheit?

A. 200

B. 250

C. 300

D. 350

15. What is the designed loading capacity for new ground ladders (except folding and attic ladders)?

A. 250 lb

B. 500 lb

C. 750 lb

D. 1,000 lb

16. In order to safely reach their objectives, aerial ladders and telescoping booms require all of the following except which choice?

A. a clear line of sight path from the turntable to the objective

B. the ability to reach over and around some obstacles

C. sufficient length

D. sufficient strength

17. An aerial ladder should be extended at least how far above a roofline?

A. 1 rung
B. 2 rungs
C. 5 rungs
D. 5 ft

18. Choose the least correct statement concerning placement of an elevating platform basket.

A. On a flat roof with no parapet, place the basket just on the roof.
B. For a moderate height parapet, place the basket with the top rail even with the top of the parapet.
C. For parapets over 6 ft high, try to find another location, or use a ladder to ascend/descend.
D. Telescoping platforms placed on the roof must first be retracted in order to clear the roof edge.

19. Which is the incorrect choice regarding the scrub area of an aerial ladder or telescoping platform?

A. The scrub area depends on the length of the device and the number of sections.
B. The greater the working height, the larger the scrub area.
C. Positioning too close to the building will reduce the scrub area on the lower floors.
D. The farther away from the building the apparatus, the larger the scrub area.

20. What is the most effective position for a rear-mounted device to achieve maximum scrub area?

A. nosed into the building
B. backed into the building
C. approximately 15° off parallel to the building
D. parallel to the building

21. **Which choice incorrectly describes where the apparatus should be spotted when no specific situation demands that an aerial device take a particular position?**

    A. Initially, stop about 15 ft past the near edge of the building.

    B. Stopping 15 ft past the near edge of the building allows the apparatus to be driven forward if the need arises.

    C. Position with the anticipated fire travel in mind.

    D. If no fire is visible, pull in line with the main entrance.

22. **Which choice below is an example of precontrol overhauling?**

    A. examining bushes under windows during the secondary search

    B. opening pipe chases behind the toilet once the fire has been knocked down

    C. opening ceilings to expose pockets of fire in the cockloft

    D. poking holes in balloon frame walls above a well-involved cellar fire

23. **Which is not one of the five senses used for overhauling?**

    A. sight

    B. hearing

    C. smell

    D. taste

24. **The 15-second/2-minute rule is a good guide for evaluating whether to open a wall or ceiling lacking any other evidence of fire. Which choice below correctly complies with this rule?**

    A. If a wall is so hot that you cannot leave your hand on it for 15 seconds, it should immediately be opened.

    B. If a gloved hand can be left on the wall for 15 seconds with no discomfort, do not open immediately.

    C. If it was decided not to open a wall immediately, another member should reevaluate after 2 minutes.

    D. If the temperature of the wall has decreased after the second evaluation, open it immediately.

25. At a serious fire in a cockloft area, where should the firefighter making the opening into the ceiling be located?

A. in the room of fire origin
B. in the most severely damaged room
C. in the least severely damaged room
D. in the doorway of the escape route

26. In which situation would hydraulic overhauling be most appropriate?

A. at a fire in an art museum
B. at a fire smoldering between a wooden lintel and a brick bearing wall above the lintel in an occupied factory
C. at a bedroom fire that has penetrated the ceiling and is involving three bays
D. at a church fire where fire has damaged the altar area

27. What is the most correct description of salvage operations?

A. Salvage is placing tarps over merchandise to protect against fire.
B. Salvage is opening up walls to check for extension.
C. Salvage is an effort made to reduce the damage to the remainder of the structure and contents.
D. Salvage is the process of claiming rights to property that has been saved from any of the fire's perils.

28. When you suspect that fire may be hiding in a partition between two rooms, what action should you take before opening up, if possible?

A. Examine both sides of the partition for the fastest, least damaging location to open.
B. Feel the wall to determine if there is smoke inside.
C. Open the wall at the top to vent any smoke trapped inside.
D. all of the above

29. How does the hoseline crew contribute to salvage?

A. by not throwing water at smoke
B. by breaking couplings outside a building to drain a line
C. by draining hoselines into bathtubs in high-rises
D. all of the above

30. Control of utilities is a simple description given to what is often a complex operation. Which safety precautions are stated incorrectly?

A. Send two members at all times, and equip them with self-contained breathing apparatus (SCBA) and a radio.

B. Send people with expertise in the area; for example, send someone with an electrical background for a plumbing problem.

C. Call for utility personnel to assist.

D. If in doubt of an action, seek further advice from experts.

31. What is the most important item affecting ladder selection on the fireground?

A. nested length

B. strength

C. electrical conductivity

D. length

32. A composite ladder is incorrectly described in which choice?

A. cannot conduct electricity

B. does not stand up well to high heat

C. has the same load-carrying capacity as aluminum

D. is only slightly heavier than aluminum

33. When planning the overhaul of a room that had just been knocked down by a hoseline, a ladder company officer made the following statements. She was least correct in which statement?

A. Plasterboard, or plaster on wire or wood lath, holds back a lot of fire.

B. Initial openings should be made near existing holes such as light fixtures.

C. If fire is found in any of these bays, the entire ceiling or wall must be opened.

D. Be especially alert for intersecting vertical openings where fire can climb up out of view.

08 FORCIBLE ENTRY

Questions

1. What factors should be included in the forcible entry size-up?

1. time of day
2. occupancy
3. type of tools available
4. direction that the door opens
5. location of the fire, victims, and door to be used
6. type of door, jamb, and locks encountered

A. 2, 3, 4, 5, and 6 only
B. 3, 4, 5, and 6 only
C. 3, 5, and 6 only
D. all of the above

2. **What factors help determine which method of forcible entry to use?**

   1. need for speed
   2. type of door and lock
   3. tools and personnel available
   4. the degree of damage that will be caused by a particular method

   A. 1 only
   B. 2 and 3 only
   C. 2, 3, and 4
   D. all of the above

3. **Which is not one of the four rules of forcible entry?**

   A. Try before you pry.
   B. Don't ignore the obvious.
   C. Always force the door with the fastest method possible.
   D. Maintain the integrity of the door.

4. **Key cylinders located more than _____ in. from the door edge indicate a potentially serious forcible-entry problem.**

   A. 1
   B. 3
   C. 6
   D. 9

5. **When using conventional forcible-entry techniques, what factor determines the mechanics of the operation and how the tools are used?**

   A. whether the door opens in or out
   B. whether the door is set in a cement wall
   C. whether the lock is a mortise type
   D. whether the lock is a rim lock

6. When forcing an inward-opening door using conventional methods with a tool with a pronounced bevel on the fork end, how should the bevel be positioned?

A. with the inner curve toward the door
B. with the outer curve toward the door
C. with the inner curve toward the floor
D. with the inner curve toward the ceiling

7. When forcing an inward-opening door using the brute force method, what is the first way to attack the door?

A. using a sledgehammer to knock the door off its hinges
B. using a sledgehammer to knock a panel out of the center of the door
C. using a sledgehammer to breach a hole in the wall next to the lock
D. using a sledgehammer to drive a Halligan tool into the gap between the door and jamb

8. When removing an inward-opening door from its hinges, which hinge should be attacked first?

A. bottom hinge first, so that venting smoke or fire goes above members as they stand to do the upper hinge
B. top hinge last, so that venting smoke or fire goes above members as they crouch to do the lower hinge
C. middle hinge first, so the door stays in place the longest time
D. top hinge first, so that venting smoke or fire goes above members as they crouch to do the lower hinge

9. An outward-opening door recessed into a wall opening should be attacked using which end of a Halligan tool?

A. the adze end, with the fork facing back across the door
B. the fork end, with the adze facing back across the door
C. the adze end, with the adze facing back across the door
D. the fork end, with the fork facing back across the door

10. The most efficient means of removing lock cylinders is with the:

A. Sunilla tool
B. K-tool
C. slide hammer
D. lock puller

11. After the key cylinder is removed, what is the first step toward opening the lock?

A. Insert the 5⁄16-in. square stock and open the mortise lock.
B. Insert the 90° key tool and open the rim lock.
C. Insert adze of the Halligan and drive the lock off the door.
D. Pick up the cylinder, examine it, and determine what key tool to use.

12. A police lock is usually found on an external door at what type of occupancy?

A. a police station
B. an apartment
C. an occupied jewelry store
D. a heavily fortified store

13. The use of through-the-lock forcible entry is indicated by all but which style of lock?

A. rim lock
B. police lock
C. Fox Lock
D. pivoting deadbolt

14. The K-tool is ineffective for pulling the cylinder of which type lock?

A. rim lock
B. police lock
C. Fox Lock
D. pivoting deadbolt

15. When using a key tool to manipulate the mechanism of a Fox Lock, the square shaft must be rotated in which direction?

A. clockwise

B. counterclockwise

C. toward the lower set of bolts at the door's edge

D. toward the upper set of bolts at the door's edge

16. Through-the-lock forcible entry should not be used under any of the following circumstances with the exception of which choice?

A. where visibility is poor

B. when heavy fire is present

C. where there is a need for speed

D. while investigating an odor of smoke

17. What is the first action a unit encountering a Mul-T-Lock door should take?

A. Use the K-tool to pull the lock cylinder.

B. Determine if the Mul-T-Lock is engaged.

C. Cut the door with the metal cutting saw and reach in and open the lock.

D. Cut the bars with the metal cutting saw and pull the door open.

18. Which of the following methods can be used to determine if a Mul-T-Lock is engaged?

A. Slide an old credit card under the door.

B. Slide a thin knife under the door.

C. Look between the door and the jamb for bars.

D. all of the above

19. Security gates include all but which type?

A. mechanical

B. electrical

C. hydraulic

D. manual

20. What is usually the key to forcing roll-up security gates?

    A. a lock puller and its key tools
    B. a metal-cutting saw or torch
    C. a knowledge of gates and their power sources
    D. a knowledge of padlocks and how to remove them

21. All but one of the following choices correctly describes high-security padlocks. Which choice does not belong?

    A. have relatively poor shear strength
    B. have a shackle of ¼-in. diameter or greater
    C. are case-hardened
    D. resist compression by bolt cutters

22. What tool is the most suited to removing padlocks located at or above shoulder height?

    A. pick
    B. duckbill
    C. power saw with aluminum oxide blade
    D. torch

23. When a roll-up gate is encountered with a padlock that is heavily shielded from tampering by a steel guard, what is the best way to attack it?

    A. Use a slide hammer to pull the lock cylinder.
    B. Use a torch to cut the lock hasp.
    C. Cut the U-channel to which it is secured.
    D. Cut the gate.

24. The American Lock 2000 series gate lock may be removed using all but which tool?

    A. duckbill
    B. power saw with aluminum oxide blade
    C. torch
    D. pipe wrench and cheater bar

25. When encountering a padlocked security gate, it is usually better to cut the padlock than the gate itself. All but one of the following are correct reasons for this statement. Which choice does not belong?

A. Cutting the gate requires a different saw blade than cutting the lock does.

B. Cutting the gate does not maximize access and ventilation.

C. It can take longer to cut the gate than to cut one or two padlocks.

D. Cutting the gate destroys it, which would not be justified by a minor fire or emergency.

26. What conditions usually require cutting open a roll-up security door?

1. Heavy fire is present behind the door.
2. Speed of water application is critical, and damage is not a concern.
3. It is an electrically operated door with the controls inside the fire area.

A. 1 only

B. 1 and 2 only

C. 1 and 3 only

D. all of the above

27. What is not an advantage of the three-cut method of cutting a roll-up security door?

A. It requires less cutting than the inverted V cut.

B. It is not hampered by wind tabs.

C. It can be used on doors recessed into a wall.

D. It guarantees clearing a larger area for access and ventilation.

28. Which choice incorrectly describes removing HUD (U.S. Department of Housing and Urban Development) sealed windows from the inside?

A. Warn personnel in the building that such windows are to be vented.

B. Drive the 2×4 braces up or down, or cut the bolts with bolt cutters.

C. Make sure someone is holding the covering, or tie a rope to it.

D. Bring the covering into the window if possible.

29. Which choice incorrectly describes removing HUD sealed windows from the outside?

A. Use a torch to cut the bolt heads off and pull the cover off.

B. Use a metal-cutting blade on a circular saw to cut the bolt heads off, and pull the cover off.

C. Use a wood-cutting blade on a circular saw to cut the outer braces and plywood around each bolt.

D. Use a wood-cutting blade on a circular saw to cut through the inner 2×4 braces.

30. Series locks are often found on commercial establishments with many exits. They can only be opened by fire personnel using which method?

A. conventional forcible entry

B. through-the-lock

C. a key from any of the other locks in the building

D. all of the above

09 VENTILATION

Questions

1. What are the two reasons for venting?

1. venting for fire
2. venting for smoke
3. venting for life
4. venting for property

A. 1 and 2
B. 1 and 3
C. 1 and 4
D. 3 and 4

2. What creates the difference between the two types of ventilation referred to in question 1?

A. the location of the ventilation
B. the style of the ventilation
C. the timing of the ventilation
D. the source of the ventilation

3. "Immediate ventilation performed to draw the fire away from the life hazard" is a description of what type of ventilation?

   A. venting for fire
   B. venting for smoke
   C. venting for life
   D. venting for property

4. "Ventilation delayed until a charged hoseline is in place" is a description of what type of ventilation?

   A. venting for life
   B. venting for fire
   C. venting for smoke
   D. venting for property

5. Where should ventilation occur to allow a hoseline to advance?

   A. at roof level
   B. in proximity to the hoseline
   C. opposite the hoseline's advance
   D. behind the hoseline

6. Horizontal ventilation for a life hazard must be accompanied by which (most correct) choice?

   A. an immediate attack on the fire
   B. an immediate rescue effort
   C. choice B or a hoseline placed to protect the victim in place
   D. choice B plus a hoseline placed to protect the victim in place

7. Which choice is not one of the factors that influence the decision to use either horizontal or vertical ventilation?

   A. life hazard
   B. size and location of the fire
   C. construction of the building
   D. effects of weather, especially humidity

8. Which choice is not an advantage of horizontal ventilation over vertical ventilation?

A. It's faster and easier to perform.
B. It's less costly to repair.
C. It's more effective at fires with heavy hot smoke.
D. It's more effective at low-heat fires.

9. After assigning the initial attack team and search teams, the IC should assign an outside ventilation (OV) team. Select the least correct statement concerning this OV team and its assignment.

A. The assignment of such a team is vital in high-security and hurricane-prone areas.
B. The OV team is responsible for an immediate survey of the inside of the structure.
C. The OV team is responsible for locating victims and the seat of the fire.
D. The OV team is responsible for creating firefighter escape routes.

10. Which fires are most likely to benefit from roof cutting?

A. low-intensity fires on lower floors
B. low-intensity fires on upper floors
C. serious fires on lower floors
D. serious fires on the top floor, attic, and cockloft

11. Mechanical ventilation is definitely called for in which of the situations below?

1. smoldering fire in stuffed chair or sofa in a windowless area
2. fire in a room that has been controlled by a sprinkler system
3. low-heat fire below grade
4. a store fire in a one-story commercial that is blowing out every window

A. 1, 2, and 3 only
B. 1 and 3 only
C. 2 and 3 only
D. all of the above

12. Mechanical ventilation involves all of the following items except which choice?

A. directing a fog stream out a window after knocking down the fire
B. venting a skylight or scuttle cover over the seat of the fire
C. placing a portable fan
D. turning the building's heating, ventilation, and air conditioning (HVAC) system to the exhaust mode

13. Which choice is not a possible drawback to venting using a fog stream?

A. very demanding in terms of personnel required
B. unnecessary water damage
C. ice hazard in freezing temperatures
D. a drain on a limited water supply

14. Which choice does not properly reflect the drawbacks of using a fan in the negative pressure (sucking) mode?

A. It requires sealing the opening around the fan, or else churning results.
B. Locating the fan in the doorway creates access and safety problems.
C. Objects being drawn into the exhaust screen block efforts.
D. Drawing combustible gases through the fan might result in ignition of the gas.

15. What factors must you take into consideration when deciding to use positive pressure ventilation?

1. location and extent of fire
2. life hazard that may be affected by fire or venting
3. availability of a hoseline
4. degree of confinement possible
5. debris that might be drawn into the fan
6. equipment available for the job, including the power supply
7. locations of exhaust openings

A. 1 2, and 3 only
B. 1, 2, 3, and 4 only
C. 1, 2, 6, and 7 only
D. all of the above

16. What is the best way to prevent mushrooming in multistory structures?

A. Ventilate all windows on the upper floors.

B. Ventilate the top of the staircase.

C. Vent directly over any vertical arteries the fire is traveling in.

D. Ventilate the fire floor thoroughly.

17. When breaking a skylight over a staircase, you must take all but which actions?

A. Thoroughly clear the entire opening.

B. Drop the entire housing through the opening.

C. Break a single, small pane first, then pause a few seconds.

D. Call on the radio, "I am taking the skylight!"

18. What style roof is this?

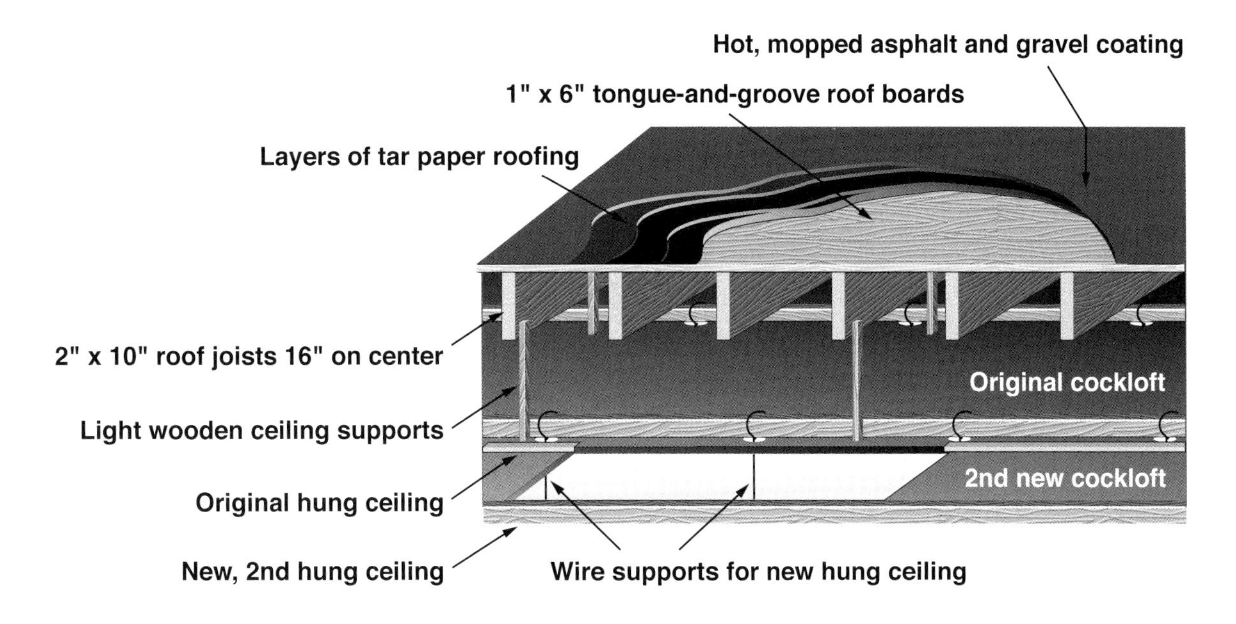

A. standard flat roof

B. inverted roof

C. rain roof

D. snow roof

19. What style roof is this?

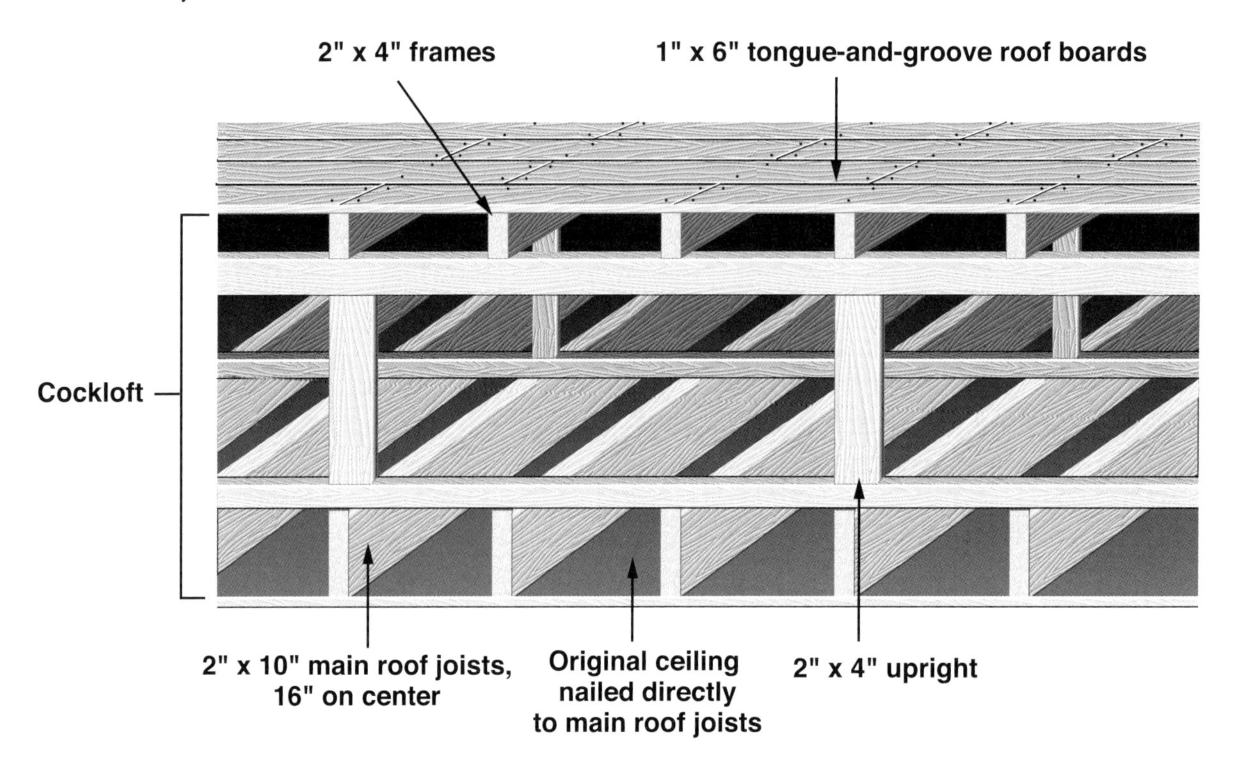

A. standard flat roof
B. inverted roof
C. rain roof
D. snow roof

20. What is the minimum commitment of personnel and equipment required to successfully vent a working fire on the top floor of a large-area flat-roof structure?

A. two people, with an axe and a pike pole or hook
B. four people with an axe, a Halligan tool, a pike pole, and a saw
C. four people with an axe, a Halligan tool, two hooks, two saws, and a portable radio
D. eight people with two axes, two Halligans, two pike poles, two saws, and two radios

21. Which statement is most correct regarding ventilation and old style or standard flat roofs?

A. They are substantial roofs, with 2×10 or 3×12 joists spaced 16 in. apart.
B. The roof deck is typically 1×6-in. tongue-and-groove boards nailed directly to the ceiling.
C. Generally, the supporting joists will burn through before the roof boards fail.
D. If a cockloft is present, it serves to insulate the roof from a fire below.

22. What is the preferred style of examination hole in a roof to determine if there is fire below, especially in daylight?

    A. quick cut
    B. kerf cut
    C. triangle cut
    D. basket cut

23. All but one of the following correctly describes the effect of wind on where ventilation holes are placed on flat roof. Which choice does not belong?

    A. It can blow fire toward escape paths.
    B. It can blow fire toward nearby exposures.
    C. It can prevent extending a previously made ventilation hole.
    D. It can prevent cutting the first holes in the best location if it is upwind of subsequent cuts.

24. What is the proper sequence of cuts to complete the 4×4-ft vent hole shown below?

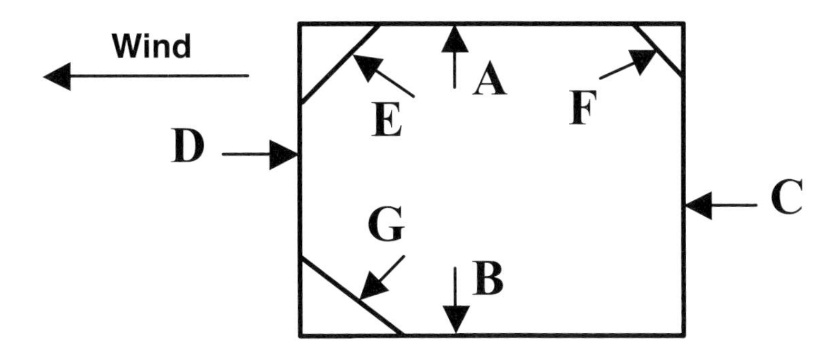

    A. C, F, A, B, D, G, E
    B. C, A, B, D, F, E, G
    C. D, E, A, G, B, C, F
    D. D, B, C, A, E, F, G

25. At a large-area flat-roof structure such as a school, supermarket, or apartment house, what size ventilation opening is recommended for a serious fire condition?

    A. 4×4 ft
    B. 6×6 ft
    C. 8×8 ft
    D. 12×12 ft

26. What is the most likely part of a standard flat roof to fail due to fire?

A. the roof trusses

B. roof boards

C. roof joists

D. roof covering

27. An inverted roof has the roof boards nailed to what element?

A. the top chord

B. the top of the roof joists

C. the bottom of the roof joists

D. a raised 2×4 framework

28. A chief officer conducting a drill on roof ventilation made the following statements regarding inverted roofs. The officer was correct in all but which statement?

A. This type roof is relatively stable, even under fire conditions.

B. It is very similar to 2×4 truss construction, in that roof failure occurs locally in sections, after a warning sag has developed.

C. Even if it does fail, it only drops down to rest on the main roof joists, which are usually intact.

D. It is not necessary to evacuate the entire roof if a serious fire is present in the cockloft.

29. A rain roof poses all of the following dangers to firefighters during a fire except which choice?

A. The presence of two cocklofts slows fire travel.

B. The added weight may have overloaded supports, precipitating collapse.

C. Rain roofs delay ventilation of the original cockloft area.

D. The presence of two cocklofts may result in conflicting estimates of a fire's intensity.

30. **While discussing new roof panels to be installed on a row of townhouses that are being built in an engine company's first-due response area, the roofing contractor tells the firefighters that the roof, built of Insulspan panels, would have the following characteristics. It should be clear that the contractor does not recognize the firefighters' problems with these panels in which statement?**

    A. These panels are up to 8 ft wide and 10 in. or more thick.
    B. They are up to 24-ft long and can span from ridgepole to bearing wall.
    C. They consist of Styrofoam bonded to OSB (oriented strand board) panels, which are noncombustible.
    D. If need be, they can be cut using a chain saw from an elevating platform.

31. **Which statement regarding a trench cut is most correct?**

    A. A trench cut is an offensive measure designed to slow fire spread to exposures.
    B. A trench cut should be made immediately at serious fires in large buildings.
    C. The best place to locate a trench is in the throat between wings of a building.
    D. The trench should be opened as soon as possible, while it is being cut.

32. **All of the following elements of an attack are needed to make a trench cut successful, except which choice?**

    A. A large vent hole must be opened over the fire.
    B. The ceiling must be fully opened directly below the trench.
    C. Defensive hoselines must be in place on the top floor as well as the roof.
    D. Additional vent holes must be cut on the safe side of the trench to ensure that fire has not gotten past the trench.

33. **Making a trench cut consists of all of the following procedures except which choice?**

    A. cutting the main vent hole
    B. locating the trench
    C. cutting inspection holes
    D. cutting the trench

34. Which statement concerning Thermopane windows, fire operations, and the dangers they pose to firefighters is least correct?

A. The newer Thermopane windows keep more heat in the building.

B. With the newer double- and triple-paned windows, you must vent just as the attack crew enters the fire area.

C. They are more difficult to vent, especially with the hose stream.

D. It is very difficult to remove the sash.

35. What is one of the greatest dangers to firefighters working on venting a roof?

A. smoke obscuring the firefighter's vision, especially in daylight

B. fire venting from a kerf cut

C. the firefighter with the power saw

D. exposure hazards that might eliminate their only escape route

36. If you are operating the saw, what is the most important action you can take when operating?

A. Watch the path the cut will take, particularly the area in front of you.

B. Stop the saw blade from spinning when it is not in a cut.

C. Cut only when the area is thoroughly clear of obstructions, such as wooden boards.

D. If you must move in a smoke condition, probe each step gently with your frontmost foot.

37. What factor should be suspected by firefighters at a fire in a jewelry store (or other high-value occupancy) in a high crime area?

A. assault by armed robbers after forcing entry

B. electrocution from theft-deterrent systems

C. steel plating added to roofs, walls, or ceilings

D. being locked inside by self-locking doors behind them

38. All of the following are critical items for the roof team to report to the IC except for which choice?

A. size and shape of the building, particularly depth

B. construction of the roof, particularly the covering

C. exposures, especially taller downwind buildings and frame buildings

D. location of parapet walls, as well as setbacks and extensions that are not visible from the street

10 SEARCH AND RESCUE

Questions

1. What are the two major phases of search that should be performed?

1. search for life
2. search for fire
3. primary search
4. secondary search

A. 1 and 2 only
B. 3 and 4 only
C. all of the above
D. none of the above

2. What is the major difference between the two phases of search?

A. time of day
B. life hazard
C. degree of fire control
D. search for fire

3. A quick search for live victims in the most likely areas they'd be, before the fire is under control, is a description of which phase?

A. search for life
B. search for fire
C. primary search
D. secondary search

4. A search performed to ensure that there is absolutely no possibility of any victims remaining in a structure is called what?

A. search for life
B. search for fire
C. primary search
D. secondary search

5. Which choice most correctly lists an advantage that thermal imaging cameras (TICs) have in fireground search?

A. A TIC resembles a video camera that shows you the actual color of the object without smoke in the way.
B. TICs can see clearly through smoke, and the human form is always clearly visible.
C. Searchers using a TIC must still perform the basics to keep themselves oriented.
D. TICs are rugged devices that can take any abuse the fire service throws at them.

6. Thermal imaging cameras (TICs) are great aids to performing a search for victims as well as search for fire. Select a true statement concerning TIC use.

A. A firefighter appears the same as a civilian victim in a TIC.
B. A firefighter using a TIC can disregard basic (slower) search techniques.
C. Firefighters using a TIC must be aware of the Superman syndrome.
D. Firefighters using handheld TICs tend to develop tunnel vision more readily than those using other types of TICs.

7. All firefighters entering a burning building must be able to perform two emergency maneuvers to allow them to operate safely while wearing the SCBA. Which statement below regarding these emergency maneuvers is incorrect?

A. The reduced profile maneuver is intended to allow a firefighter to escape from a tight spot, such as between the studs of a partition wall.

B. To reduce your profile, you align the SCBA cylinder with your body, leaving the facepiece in place, but taking the harness off entirely.

C. Begin the reduced profile by fully extending the right shoulder strap and removing your right arm from the harness.

D. Twist the mask assembly to the left, far enough to pass the obstruction.

8. The primary search of the fire area is usually best begun where?

A. moving quickly toward the fire, then working back toward the door or exit

B. inside the fire room

C. around the perimeter of the entire building

D. at the door to the area, then working in toward the fire

9. Where should you begin searching when entering the area directly above the fire?

A. moving quickly toward the fire, then working back toward the door

B. inside the fire room

C. around the perimeter of the entire building

D. at the door to the area, then working in toward the fire

10. To which areas should you give primary attention when searching residential structures?

A. bedrooms

B. paths of egress

C. stairways

D. all of the above

11. What methods can you use to improve conditions in the search area while performing the primary search?

1. Close doors between yourself and the fire.
2. Vent windows in the room being searched.
3. Close the door to the fire room if possible.

A. 1 only
B. 2 only
C. 1 and 3 only
D. all of the above

12. Recognizing a child's crib in total darkness takes some practice and the ability to recognize what you are feeling. Which statement below most accurately describes most cribs?

A. usually have high legs, with low sides
B. unusually have low mattress level, confined on at least one end or side
C. usually have low legs, confined on at least one end or side
D. usually have high legs, with barred sides

13. Recognizing a bunk bed in heavy smoke and darkness may mean discovering a victim that might otherwise be overlooked until it is too late. Which description below most accurately reflects bunk bed design?

A. usually has high legs, with low sides
B. unusually low mattress level, confined on at least one end or side
C. usually has low legs, confined on at least one end or side
D. usually has high legs, barred sides

14. As the incident commander, your units have knocked down a heavy fire on the ground floor of a three-story, brick and wood joist row house, with two apartments per floor. What is the desirable minimum number of personnel you should assign for secondary search purposes?

A. 2
B. 6
C. 12
D. 24

15. **Which statement about search techniques does not properly reflect good tactics?**

A. Gathering information is a critical part of search. If you know where a victim is located, get there by fastest route available.
B. Search of the immediate seat of the fire area may be accomplished by the nozzle team.
C. Bedrooms above the fire are often best reached by vent, enter, search (VES).
D. The area immediately below windows must be searched because evidence may be found there.

16. **Which choice below is least correct concerning secondary search techniques?**

A. The secondary search may be combined with postcontrol overhauling.
B. Before moving objects such as draperies and chairs, thoroughly search them.
C. It is usually best to have different teams conduct the primary and secondary searches of an area.
D. Be sure to examine the perimeter, including rooftops and setbacks that people may jump to.

17. **To perform a guide rope–assisted search, what items should members have on hand before entering the search area?**

1. two-way radios
2. SCBA, PASS device, and flashlight for each member
3. search rope (200 ft or more of light line)
4. forcible-entry tools and TIC
5. large floodlight at the entrance

A. 1 and 2 only
B. 1, 2, and 4 only
C. 2 only
D. all of the above

18. **When using a guide rope, is it necessary for all members to remain in constant contact with the guide rope?**

A. yes
B. no

19. The reduced profile and the emergency escape procedures allow members wearing SCBA to free themselves from entrapments. Which statement below correctly describes these procedures?

A. To reduce your profile, remove your right shoulder strap and spin the bottle to the right until it is in line with your body.

B. The emergency escape procedure involves removing the SCBA entirely from your back, allowing you to pass obstructions while entering a fire area.

C. Both the reduced profile and emergency escape can be performed with the facepiece in place on your face.

D. An emergency escape starts with removing your right shoulder strap, then waist and chest straps, keeping your right hand on your right shoulder strap the entire time.

20. Firefighter safety should be the highest priority during any operation. Which statement below incorrectly describes safety considerations during a search?

A. Each team should be equipped with at least one powerful flashlight.

B. When making tool assignments, ensure that each interior search team is equipped with forcible-entry tools.

C. A pike pole is very useful for venting windows as well as probing with the blunt end under beds and other furniture.

D. Each team should be equipped with at least one portable radio.

PART II

SPECIFIC FIRE SITUATIONS

11 FIREFIGHTER SURVIVAL

Questions

1. **All but one of the following are actions to be taken to reduce the firefighter mortality rate. Which choice is not an immediate step?**

    A. Improve hazard awareness and recognition training.

    B. Provide firefighters with emergency escape capability.

    C. Provide firefighters with additional escape routes.

    D. Provide firefighter rescue teams.

2. **Which choice is not one of the three main rules of survival?**

    A. Always have a charged hoseline while searching.

    B. Never get into a position where you are depending on someone else to come and get you out.

    C. Always know where your escape route is.

    D. Always know where your second escape route is.

3. **Hazard awareness is critical to firefighter safety. Which of the following questions is not one of those outlined in the firefighters survival survey?**

    A. What is the occupancy?

    B. What is the time of day?

    C. Where is the fire?

    D. How do we get in?

4. A company officer discussing the firefighters survival survey with the company members made the following comments concerning "What is happening to the building?":

1. You must constantly be evaluating the potential for flashover.
2. You must be evaluating the potential for a backdraft.
3. You must be aware of fire traveling in voids around you.
4. You must be aware of the size of the floor area above you.
5. You must monitor the stability of the building.

This officer was correct in which choice(s)?

A. 1, 2, 3, and 5
B. 3, 4, and 5
C. 5 only
D. all of the above

5. If you find yourself entangled in cable TV wires while searching, which choice most correctly states a recommended action to take?

A. Remove the SCBA harness to untangle it from the wires.
B. Inform the incident commander of the difficulty.
C. Try backing up, standing up, and proceeding forward.
D. Perform the Heimlich maneuver.

6. Which choice would not be a reason for performing the emergency escape maneuver?

A. to free yourself from a serious entanglement
B. to permit you to escape between the studs of a wall you breached
C. to allow a member to crawl out from beneath a beam blocking his or her exit
D. to allow a nozzle operator to enter an extremely narrow space while looking for fire

7. If firefighters find themselves cut off by rapidly spreading fire, they should take all but one of the following actions. Which choice does not belong?

A. Stay calm and stay put to conserve air.
B. Find an area of refuge, then close doors between you and the fire.
C. Call for help by voice, radio, and PASS.
D. Locate any exit, including breaching a wall or a window.

8. A firefighter that finds himself lost and running out of air must escape to fresh air very quickly. Four members in such a predicament each performed one of the following acts. Which member acted unwisely?

1. A member broke a window and climbed headfirst down a ladder to avoid fire that was venting out of the window over his head.
2. One member kicked a door open with her feet to escape.
3. A third member used a Halligan tool to breach a wall through to an adjoining apartment.
4. The fourth member did an emergency bailout from a window by wrapping a personal rope around his chest.

A. 2 only
B. 1, 2, and 4 only
C. all of the above
D. none of the above

9. At times, a firefighter may find herself lost in a smoke-filled building where there is no danger of being overrun by fire, but where a life-threatening atmosphere is present. This may occur in a building that is sprinklered, but where the sprinklers cannot fully extinguish the blaze. Which choice is an incorrect action for the lost firefighter to take?

A. Stay calm and stay put to conserve air.
B. Call for help by voice, radio, and PASS.
C. If she must seek an exit, walk with her light off, so she can see other lights.
D. Use all means at her disposal to attract attention to herself.

10. An officer instructing members on the technique of the emergency body wrap with a personal rope cited the following statements as being critical to success. This officer was incorrect in which statement?

A. When descending, don't allow your hands to spread farther than about shoulder width apart.
B. Remove all slack from the rope just prior to exit.
C. Never attempt this maneuver without a turnout coat, gloves, and a second member belaying the member descending.
D. An NFPA-compliant one-person rope should be used for this escape.

11. One simple method of establishing accountability for the members of a career department is the written riding list. Select the incorrect statement regarding the riding list.

A. The list must be prepared at the start of the tour, and it shows all members working in the unit.

B. The list must be updated to reflect any changes, such as if a member leaves sick.

C. The list should be prepared in triplicate.

D. The officer keeps the original for roll calls and gives the duplicate to the command post at large-scale operations.

12. What is the most important piece of equipment on the fireground once a member has been reported missing?

A. a riding list

B. a charged hoseline

C. portable radios

D. a PASS device

13. Roll calls should be conducted in all but which of the following situations?

A. Mayday has been transmitted.

B. PASS alarm ceases operating.

C. Structural collapse has occurred.

D. Sudden fire extension or orders have been given to withdraw from a building.

14. None of the following situations would justify transmitting a Mayday except which choice?

A. A member is hung up on wires or other obstructions that will require 3 seconds to clear.

B. An injury to a member has occurred.

C. An unconscious person has been discovered on the fire floor.

D. Collapse is imminent and will endanger members.

15. An incident commander who is aware of a Mayday must get as much information about the situation as possible. Which choice most correctly identifies the items the IC needs to ask about?

A. What is the rank of the person who transmitted the Mayday and the nature of the emergency?

B. How did the member enter: front door, rear, side, or fire escape?

C. Can the member see any fire or smoke?

D. How long has the member been trapped?

16. What are the prerequisites to a successful rapid intervention team (RIT) operation?

A. people, planning, and training
B. people, policies, and training
C. people, planning, tools, and techniques
D. people, policies, tools, and techniques

17. According to the author, what is the minimum number of people needed to rescue a downed firefighter?

A. two
B. three
C. four
D. six

18. Which is the best choice of personnel for RIT duties?

A. an engine company
B. a ladder company
C. an ambulance crew
D. a specially trained ladder company

19. Select a true statement regarding a RIT:

A. A RIT should not perform any functions until a firefighter is trapped.
B. A RIT should always remain in the vicinity of the command post.
C. A RIT should never be involved in using a handline.
D. A RIT should automatically be dispatched on the report of a working fire.

20. Select an untrue statement concerning the operation of a RIT.

A. A RIT might be used as an attack unit if the incident escalates.
B. A RIT should never be used for overhaul.
C. A RIT that has been deployed for rescue duties must be replaced.
D. A RIT must conduct a size-up and determine the location of operating units.

21. **All members of a RIT should step off their apparatus with all of the following equipment except for which choice?**

    A. a large handlight, preferably on a sling for hands-free operation
    B. a good, sharp knife
    C. an NFPA-compliant personal rope
    D. SCBA, with an activated PASS and a spare bottle

22. **A RIT officer should ensure that his six-person team report in with the proper tools. Which choice is the most correct listing?**

    A. a set of forcible-entry tools
    B. a power saw, either wood- or metal-cutting, depending on size-up
    C. a 200-ft search guide rope for each member
    D. a spare mask for each member
    E. all of the above

23. **At the scene, a RIT should attempt to locate all but which of the following items?**

    A. a personal rope and harness
    B. a copy of the building's floor plan
    C. a copy of the command chart, indicating which units are operating and where
    D. a hoseline that can be committed for RIT use, and its supply

24. **As part of the RIT size-up, the team should have which of the following?**

    A. a radio report while en route giving the building size, construction, occupancy, and fire location
    B. someone monitoring tactical and command channels for urgent or Mayday messages and the progress of the operations
    C. at the scene, the building size-up, occupancy, location and extent of fire, location of units operating, and routes of access they have used
    D. all of the above

25. As soon as the first report of a downed firefighter is received, all but one of the following actions should be taken. Which choice does not belong?

A. A BLS (basic life support) ambulance should be called if none is on the scene.

B. Additional firefighting personnel should be called.

C. A protective hoseline and a spare mask should be brought to the vicinity of the operation.

D. A BVM (bag valve mask) resuscitator with oxygen should be brought to the vicinity, even if there is no apparent need for it.

26. All of the following methods may be helpful in locating an unconscious firefighter except which choice?

A. feedback-assisted rescue

B. pager-assisted rescue

C. portable radio distress signal

D. PASS device

27. Which of the following is not a drawback of feedback-assisted rescue?

1. The missing member must be radio equipped.
2. It takes time to implement.
3. It requires alternate means of communicating.
4. It ties up what may be limited radio channels.
5. It requires members to be recognized as being in trouble.

A. 3 only

B. 3 and 4 only

C. all of the above

D. none of the above

28. Team search differs from using a rope merely as a guideline. Select the incorrect statement regarding team search operations.

A. During team search, all members must be connected to the guide rope, either directly or by a short rope.

B. A definite distance limit is set for the team to operate.

C. A control person remains outside to monitor the team.

D. A rescue team must be standing by, immediately available at the entrance, to assist as needed.

29. When discussing team search, a senior firefighter made the following statements. The firefighter was correct in which choice(s)?

1. Team search is not generally useful for locating civilians in a fire.
2. Team search is a RIT skill that may locate trapped firefighters.
3. The fan search works best in large, wide-open spaces as a fast way to search.
4. A means of keeping track of how far into the area you have traveled is needed.
5. The T search is more thorough than the fan search, but slower.

A. all of the above

B. 1, 2, 3, and 4 only

C. 1, 2, 4, and 5 only

D. 1, 2, and 4 only

30. What precaution is necessary when changing direction while using a search or guide rope?

A. The rope must be secured outside the building at all times.

B. The rope must be secured at each floor of a multistory building.

C. The rope must be secured at each change of direction.

D. all of the above

31. Several factors make the rescue of a trapped firefighter different from the rescue of a civilian. Which of the following statements do not apply?

A. All firefighters are full-grown adults; there are no 55-lb firefighters.

B. Turnout clothing and SCBA add weight and could get hung up on obstacles.

C. There is added psychological stress involved in firefighter rescue.

D. The trapped firefighter can at least help with her own rescue.

32. A rescue company officer giving a drill on trapped firefighter rescue made the following statements. This officer was most correct in which choice?

1. If fire conditions and logistics permit, use positive pressure fans to provide clean air to the trapped members and rescuers.
2. Begin multiple avenues of approach to trapped members.
3. A clear chain of command is needed at these incidents.

A. 1 and 2 only
B. 2 and 3 only
C. all of the above
D. none of the above

33. The incident commander at an incident where a trapped or missing firefighter is reported needs a clear plan of action if the situation is to be rectified in time to save the firefighter in distress. Which of the choices below is not listed as one of the IC's duties in such a situation?

A. Take control of the fireground radio traffic and gather needed information.
B. Assign appropriate resources to remove the member, and announce the identity of the units assigned to this task on fireground radio frequencies.
C. Direct and control the removal efforts personally to avoid conflict among subordinates.
D. Reevaluate and modify the firefighting strategy as necessary.

34. Members performing RIT duty must be properly trained and equipped. Which statement below would be an incorrect statement regarding this duty?

A. If you are first to discover a trapped member, call for assistance immediately.
B. Never share your mask with a trapped member.
C. Be totally familiar with all types of SCBA used in your department and responding mutual aid units.
D. If you have dragged a victim out of immediate danger, begin basic life support and await aid.

35. When an unconscious firefighter must be removed from danger immediately, any of a variety of methods may be used. Select a true statement regarding such movement.

A. The preferred drag is the feet-first method.

B. Converting the downed member's SCBA into a harness assists removal efforts.

C. If the SCBA keeps getting hung up on obstructions, take it off and place it on the member's chest.

D. If the victim is entangled in an obstacle, the rescuer will have to feel all around the victim for the entanglement and may have to forcibly pull the victim free.

36. If a firefighter has to be removed up a staircase to safety, which choice below properly describes a technique that is recommended?

A. Two firefighters can carry the member up a tight staircase by working side by side at the victim's shoulders.

B. If the victim is heavy, make the SCBA a harness and run a rope through the waist belt, then have rescuers at the head of the stairs pull while one rescuer guides the victim's head and shoulders.

C. The rescuer with the victim must closely control the hauling, signaling the others at the top of the stair when to haul.

D. Another approach uses a backboard. Run one end of the rope up the stairs, secure it to the backboard, and use it to pull the backboard up.

37. Choose the correct statement regarding the removal of a large, unconscious firefighter out a window onto a portable ladder.

A. It may be possible to use tools to create a ramp to help raise the victim to the windowsill.

B. If this rescuer can get under the victim, the only other help needed is one person on the ladder.

C. If the victim cannot be positioned on a rescuer's back, then the best choice is to manually lift the victim up onto the sill with three or four people.

D. Ladder rescue can be made safer by using a stronger ladder.

38. At times it may be necessary to use a ladder as a high point anchor to allow members outside on the ground to assist firefighters inside an upper floor in lifting and hauling a heavy person out of a window, particularly if the windowsill is high above the floor. Select an incorrect statement concerning this procedure.

A. A rope secured with the rescue knot and slippery hitch around the victim's chest is the preferred method.

B. The rope leads out the window, over a rung of the ladder, which is at least 4 ft higher than the sill or even with the top of the window, then to the ground.

C. On signal, three or four people pull from below while members in the room guide the victim to the sill, then the people on the ground lower the member to safety.

D. The handcuff knot may also be used to attach the rope to the victim.

39. If extreme conditions exist, it might be necessary to haul an unconscious person up vertically through a hole in the floor. Choose the incorrect statement describing this maneuver.

A. Another member must descend to the victim's location.

B. The middle of a length of rope is lowered to this member, who secures the victim's wrists or ankles with a handcuff knot.

C. If possible, lower a second rope, also by the middle, and place another handcuff knot on the other extremities.

D. Once secured, on signal, four rescuers should haul the victim vertically through the opening.

40. Hauling a victim vertically by rope has several advantages. Select the least correct choice.

A. By hoisting on the extremities, the arms are always overhead, reducing the body's profile.

B. The shoulders are normally the widest part of the body.

C. Hoisting via the arms serves to anatomically align the spine.

D. If conditions permit, place a backboard at the mouth of the opening and secure the victim to it for transport.

12 OPERATIONS IN LIGHTWEIGHT BUILDINGS

Questions

1. Of all the types of buildings that firefighters may have to operate in, those built with any of the lightweight construction elements are the most dangerous. This includes all but which of the following choices?

   A. trusses and bar joists
   B. C-joists
   C. sawn joists
   D. plywood I-beams

2. What course of action is needed when dealing with lightweight buildings?

   A. Use tactics that allow us to complete a primary search and achieve fire control.
   B. Ignore the issue, as it is a code enforcement one that doesn't affect firefighters.
   C. Change our procedures when dealing with lightweight buildings.
   D. Ensure that building codes do not permit such construction.

3. What two conditions exist that make operations more dangerous in today's buildings?

   A. lightweight construction and trusses
   B. trusses and plywood I-beams
   C. lightweight construction and new buildings
   D. lightweight construction and renovations

4. **Select the least correct statement regarding building construction from the choices below.**

   A. Virtually any type of building or occupancy can be found with either standard or lightweight construction.

   B. The primary difference between standard and lightweight construction lies in the mass of the elements that carry the loads.

   C. Older Class 4 or Class 5 buildings have relatively large members such as 2×10 joists or heavy, steel I-beams.

   D. Using a standard wood beam, about the maximum distance that can be spanned is 25 ft.

5. **What is the most dangerous type of roof for a firefighter to work on or under?**

   A. standard flat roof

   B. inverted roof

   C. peaked roof

   D. truss roof

6. **Which of the following is not one of the major design flaws of all trusses, from a firefighting standpoint?**

   A. The lack of mass compared to standard beams means earlier failure.

   B. The failure of a single part does not destroy the entire truss.

   C. The open space within the truss allows rapid fire spread throughout the area.

   D. The connectors often fail or speed the failure of the truss members.

7. **What is not one of the three elements of any truss?**

   A. the top chord

   B. the middle chord

   C. the bottom chord

   D. the connectors

8. **Select the most correct statement regarding how a truss transmits its load to other elements.**

   A. The top chord is in compression.

   B. The bottom chord is in compression.

   C. The middle chord is in compression.

   D. The connectors are in neutral.

9. What is the deadliest form of roof construction per incident of all current styles?

A. bowstring truss
B. 2×4 gusset plate trusses
C. bar joist roof
D. plywood I-beams

10. What factors most correctly serve to indicate a bowstring truss?

1. classic humpback design
2. occupancy requiring large, open floor spaces without columns
3. occupancy used on bowling alleys, skating rinks, garages, etc.
4. a particularly deadly variety of bowstring truss that has hip rafters spanning two trusses

A. 1 only
B. 1 and 2 only
C. 1, 2, and 3 only
D. all of the above

11. A dangerous relative of the bowstring truss is the flat timber truss. Select the element of construction that makes flat timber trusses even more dangerous than bowstring trusses.

A. They are built of 2×8 or 2×10 top and bottom chords.
B. They are spaced approximately 20 ft apart.
C. The collapse of just one truss will create a 40-ft-wide hole.
D. Their shape offers no clue to their presence from roof level or below a ceiling.

12. What is the biggest drawback of lightweight (2×4-in.) trusses?

A. their all-wood construction
B. their lack of mass to resist fire
C. their extreme size
D. their parallel chord design

13. Many different styles of lightweight trusses are available, using short pieces of lumber joined together by various metal connectors. What danger do firefighting operations pose to all flat chord trusses?

A. The corrosion of the metal fasteners hastens collapse.
B. The small wooden elements take longer to burn through.
C. Hoseline operation tends to blow the connectors off.
D. If power saws are used to cut the deck, they may sever the top chord, precipitating collapse.

14. What problem is common to all open-web type trusses, both wood and steel?

A. They create large open spaces called a truss loft on each level.
B. The truss loft acts as an insulator against fire spread.
C. The web members are all loaded in compression.
D. The chords are all loaded in compression.

15. What factor makes plywood I-beams particularly dangerous to firefighters?

A. Fire spreads extremely slowly throughout the truss loft.
B. The burn-through time of ¾-in. plywood webs is extremely fast.
C. Plywood I-beams show no noticeable sag before they suddenly fail, dropping everything above into the fire.
D. Tremendous fire is present on the floor above the fire.

16. Which choice below correctly describes metal deck roof construction?

A. A very common residential roof involves bar joists supporting a corrugated roof deck.
B. Bar joists are formed by welding pieces of angle iron together.
C. Another style involves a lightweight steel I-beam with cutouts in the web.
D. The main drawback of all metal roof designs is the difficulty of cooling the roof with a hose stream due to its unique shape.

17. Metal deck roofs are incorrectly described in which choice below?

A. The roof decking consists of thin steel sheets 3 ft wide and 20 ft long.
B. The roof decking is covered with a layer of hot tar.
C. On top of the tar is a layer of insulation, usually mineral wool.
D. This insulation is then covered with another layer of tar as a water sealant.

18. Select the incorrect statement regarding fires in buildings with metal deck and bar joist roofs.

A. The primary hazard to firefighters is falling into the ventilation hole while they are cutting it.

B. The second danger, collapse of the roof, is greatest to firefighters on the roof.

C. The third problem with these roofs is the metal deck roof itself burning.

D. It may be necessary to cut sidewalls of these buildings to provide adequate ventilation.

19. Which statement regarding firefighting operations at buildings with metal deck roofs is least correct?

A. If heavy fire exposes the roof, collapse should be expected in 5 minutes or less.

B. If 50% of the building is heavily involved, collapse may well be inevitable.

C. If smaller areas are involved, it may be possible to cut the roof directly over the fire.

D. Use a large hoseline to cool the steel while the roof team cuts the deck.

20. A fire officer training new members of his crew made the following statements concerning metal deck roof fires. The officer was most correct in which choice?

1. A metal deck roof fire can occur with any major involvement of the stock below the roof.
2. The metal deck acts like a gigantic frying pan when heated, distilling flammable gases from the tar.
3. The result is a self-accelerating reaction that can spread quite rapidly.
4. The reach and cooling power of 2½-in. lines are helpful.
5. It may be necessary to cut through the insulation to fully extinguish the fire.
6. Metal deck roof fires can occur in fully sprinklered buildings.

A. 1, 2, 3, and 6 only

B. 1, 2, 4, 5, and 6 only

C. 1, 2, 3, 4, and 6 only

D. all of the above

21. Which statement regarding lightweight construction is least correct?

A. Lightweight construction can be found in any occupancy or building type.
B. Lightweight construction in new buildings is more dangerous than in older buildings.
C. Lightweight buildings require different tactics than older, heavier styles.
D. Altered buildings kill firefighters; thus documenting alterations is critical to firefighter safety.

22. Which statement concerning firefighting operations in lightweight buildings is least accurate?

A. When heavy fire is present in unoccupied buildings, roof ventilation should not be conducted.
B. When heavy fire is present in unoccupied buildings, aggressive interior operations should not be conducted.
C. When heavy fire is present in occupied buildings, roof ventilation should not be conducted.
D. When heavy fire is present in occupied buildings, aggressive interior operations should never be conducted.

23. A chief officer conducting a training session about tactics at lightweight private dwellings made the following statements. He should be demoted since he was incorrect in the comments found in which choice?

1. A key element in the decision-making process is the potential for saving lives.
2. If it seems that victims are savable, conditions may permit a rapid entry and rescue.
3. The safest, most direct approach may involve exterior hose streams.
4. Vent, enter, search uses an exterior approach to savable victims.
5. If there are no savable victims, do not endanger the lives of firefighters.

A. All of the above choices are improper.
B. Only 1, 2, 4, and 5 are improper.
C. Only 1, 2, 4, and 5 are correct.
D. All of the above are correct.

24. Which choice below correctly identifies a tactic that would be improper at a building that is built with lightweight construction and no savable victims?

A. Firefighting should proceed quite carefully and from a distance.

B. Use the reach of the stream to extinguish fire well before allowing members to operate directly over or below damaged trusses.

C. Opening up the building's sidewalls and directing a stream into the truss loft may be very useful.

D. Aerial ladders are the tool of choice for these operations.

25. Regarding actions to take at a fire involving a lightweight structure where a decision has been made to enter the structure, which choice below is most correctly stated?

A. After the fire has been knocked down, place the emphasis on assessing the structure.

B. Proceed quickly, providing maximum ventilation and lighting.

C. Do not operate in the danger area until the walls have been opened.

D. Shoring should be installed all around the hazard area.

13 BELOW-GRADE FIRES: BASEMENTS AND SUBBASEMENTS, CELLARS AND SUBCELLARS, AND CRAWL SPACES

1. **The disadvantages of using a cellar pipe on a severe cellar fire include all but which choice?**

   A. It requires constant staffing to operate.

   B. Operation area maybe untenable.

   C. It only applies water in one or two directions at a time.

   D. It has a limited reach of stream, only 15–20 ft.

2. **All of the following are prerequisites for high-expansion foam to successfully darken down a fire. Which choice is not correct?**

   A. The floor above the fire must remain tenable.

   B. Sufficient volume of foam per minute must be applied.

   C. Foam must be able to reach the seat of the fire.

   D. The fire must involve Class A materials.

3. **Members entering a high-expansion foam blanket should know that all of the following are true except for one choice. Which is incorrect?**

   A. The entry should only be made for lifesaving purposes.

   B. All members shall be properly equipped.

   C. Electrical supply to the area should be shut off.

   D. A rapid intervention team must be in place at the point of entry.

4. **A firefighter checking for extension over a serious basement fire gets to an outside wall, removes the baseboard molding, opens a hole through the wall, and discovers a 2×4 lying flat along the floor. What report should this member send to the incident commander regarding the building's construction?**

   A. The wall is platform framed.
   B. The wall is balloon framed.
   C. The wall is braced framed.
   D. The wall is not yet opened.

5. **Once fire is found spreading up a balloon frame wall, speed is of the essence. Which of the following is a good action to take in this event?**

   A. If high heat or fire is present, drive the hose stream up the bay.
   B. Position one hoseline on the top floor and one in the cellar.
   C. Quickly expose each bay for its entire length, especially bays above and below windows.
   D. Do not bother with roof ventilation if the fire is in the cellar.

6. **Which of the following conditions would not likely result in a cellar fire in a taxpayer that could destroy the entire building?**

   A. heavy fire loading
   B. maze-like storage conditions
   C. lack of quick ventilation
   D. presence of a sprinkler system

7. **When describing fire tactics to be used at a heavy cellar fire in a store, a company officer made the following comments. Which correctly describes an action that should be avoided in such situations?**

   A. A fast, high-volume attack should be made in conjunction with cutting holes in the first floor.
   B. A 2½-in. line equipped with a nozzle set at an approximately 45° fog pattern should be used to create large quantities of steam.
   C. A second 2½-in. line should be positioned at the base of the cellar stair to protect the first line.
   D. A 4×4-ft hole (minimum) should be cut near the front show windows, even though the fire is in the center of the cellar.

8. How much and what size hose should be ready at the front door of a 75-ft deep taxpayer before advancing into a cellar fire, assuming the cellar entrance is in the rear of the store?

A. 75 ft. of 2½-in.
B. 150 ft of 2½-in.
C. 200 ft of 2½-in.
D. 250 ft of 2½-in.

9. A ladder company captain was lecturing the members of an engine company regarding store fire tactics and made the following comments: "When the first line is being advanced through the cellar and encounters a high heat condition, the line should be operated on a wide fog to darken down fire and protect the members on the line." The captain was ________________.

A. Correct. The high heat is an indicator that the area is about to backdraft.
B. Correct. The use of the fog stream is critical to firefighter safety in this situation.
C. Incorrect. The high heat is common in cellar fires with their low ceilings.
D. Incorrect. The use of a fog stream here will likely burn the firefighters.

10. What precautions are necessary to protect members advancing a handline deep into the cellar of a taxpayer?

1. A second line should be positioned in the cellar to protect their escape.
2. A member should be positioned at the top of the cellar stair to warn members of fire behind them or on first floor.
3. The line must always be charged before going down the stair.

A. all of the above
B. 1 only
C. 2 only
D. 1 and 3 only

11. At a cellar fire in a taxpayer, the incident commander gave the following orders. She is correct in which choices?

1. Examine all voids and pipe chases for extension.
2. Position hoseline to cut off extension.
3. Cut the first floor just inside the show windows for ventilation.
4. Prepare distributors or cellar pipes, in the event handline attack fails.

5. After the hole in the first floor has been completed, use a positive pressure ventilation fan to blow air into the hole.
6. Set up a fog line inside the first floor behind the vent hole to blow smoke, fire, and gases out the front window.

A. all of the above
B. 1, 2, and 3 only
C. 1, 2, 3, and 4 only
D. 1, 2, 3, 4, and 6 only

12. Which choice is not one of the more serious threats to firefighters from a cellar fire in some taxpayers?

A. terrazzo floor
B. quarry tile floor
C. reinforced concrete floor
D. unreinforced concrete floor

13. A 4-in.-thick terrazzo floor that covers an area 10 ft wide and 10 ft long adds how much weight to the floor supports?

A. 5,000 lb
B. 7,500 lb
C. 10,000 lb
D. 15,000 lb

14 PRIVATE DWELLINGS

Questions

1. **Fires in one- and two-family homes account for most nearly what percentage of civilian fire deaths in the United States each year?**

    A. 50
    B. 60
    C. 70
    D. 85

2. **Select the most correct statement regarding fires and firefighting operations in private dwellings.**

    A. The majority fires in private dwellings begin below the first floor.
    B. The cellar often contains the heating plant, electric service, water heater, and cooking equipment.
    C. Quite a lot of cigarette smoking occurs in the below-grade areas.
    D. Upper floors normally contain sleeping areas and must be the primary target of VES attempts 24 hours a day.

3. Place the following firefighting activities in the proper sequence:

    1. Secure water supply.
    2. Perform VES of all areas.
    3. Stretch attack line to protect bedroom exits.
    4. Extinguish fire.

The correct sequence is:

A. 1, 2, 3, 4
B. 1, 3, 4, 2
C. 1, 2, 4, 3
D. 1, 3, 2, 4

4. What is the second most serious defect in a multistory private home?

A. lack of code requirements for interior finishes
B. lack of an enclosed stairway
C. large number of occupants
D. small room size

5. Which choice most correctly describes the most critical task of the first hoseline at any private dwelling fire?

A. Extinguish the fire.
B. Get between the fire and the staircase.
C. Get between the fire and the bedrooms.
D. Get between the fire and the living room.

6. An engine company captain giving a drill on private dwelling fires made the following statements. His lack of experience in private dwellings is most accurately reflected in which statement?

1. In two-story or higher homes the first line should be positioned to get water immediately on the fire.
2. Usually the best place to achieve this is through the rear or side door.
3. It is a good idea to stretch a second line to cover the floor above.
4. A third line should be stretched to cover any exposure hazards.

A. 1 only
B. 1 and 2 only
C. 1, 2, and 3 only
D. all of the above

7. When planning water supply requirements for a well-involved two-story home without any exposure hazards, you should plan for about how many gpm?

A. 100–200 gpm
B. 200–400 gpm
C. 400–600 gpm
D. 600–800 gpm

8. When discussing the advantages of inline pumping at an advanced private dwelling fire, a chief officer made the following statements. She was least correct in which statement?

A. Using inline pumping, a preconnected line can be charged as soon as the chauffeur gets out of the cab.
B. A 500-gallon booster tank can supply a 1¾-in. line for at least 5 minutes.
C. A 500-gallon booster tank will knock down the vast majority of most private dwelling fires.
D. The proper position for the first engine using an inline supply is just past the fire building.

9. **The same chief in question 8 continued with the following comments regarding private dwellings and inline pumping. Again, she showed her lack of experience in engine operations in which statement?**

    A. Inline pumping permits the fastest water application possible, using preconnected lines off booster tank.

    B. Inline pumping ensures a continuous water supply.

    C. Inline pumping makes a master stream available in front of the building.

    D. A spot should be left for an aerial device, which is not normally required at upper floor fires.

10. **When operating from a 500-gallon booster tank, how long can you maintain an effective fire stream through a 1¾-in. line at its designed flow?**

    A. 1½ minutes

    B. 2 minutes

    C. 2½ minutes

    D. 3 minutes

11. **Fires on the exterior of a structure can be just as dangerous and destructive as those confined to the inside, especially if combustible siding, either asphalt or vinyl, is involved. Choose the most correct statement regarding these situations.**

    A. In most cases, the first line should be an exposure line.

    B. The first hoseline should generally be placed to protect the interior stair, if present.

    C. The first interior line must be brought to the attic to check extension there.

    D. Depending on the ability to quickly place a third line, it may be best to hit the exterior fire as soon as the second line has been charged.

12. **Roof ventilation on a private dwelling would be justified in all but which of the following cases?**

    A. fire in an attic

    B. fire in the cellar of a platform frame house

    C. fire in the cellar of a balloon frame house

    D. fire that extends from the first floor via combustible siding to involve the eaves

13. **Arrange the following types of roof sheathing in order of the most stable to least stable:**

    1. tongue-and-groove boards

2. furring strips
3. older plywood
4. newer fire retardant plywood that has decayed.

    A. 1, 2, 3, 4
    B. 3, 1, 2, 4
    C. 1, 3, 2, 4
    D. 3, 4, 1, 2

14. A newly promoted lieutenant decided to show his crew how much he knew about roof operations. He made the following statements about the dangers that hard roof coverings pose to firefighters. He was correct in which choice?

A. Hard roof coverings are extremely slippery when wet.
B. Loose tiles or slate pose a falling object hazard to firefighters cutting them.
C. Slate compounds the fall hazard to firefighters working below them.
D. Spanish tile is best cut with a wood-cutting saw.

15. What factors should indicate the correct location of a ventilation opening on a peaked roof?

1. wind direction
2. fire location observed en route to the roof
3. visible hot spots (e.g., steam, bubbling tar, and melting snow)
4. obstructions on the roof (e.g., solar panels)

    A. 1 only
    B. 1 and 2 only
    C. 1, 2, and 3 only
    D. all of the above

16. Which choice below correctly describes roof-cutting operations on a peaked roof private dwelling?

A. On roofs of low pitch (under 45°) a power saw may be used to make the cut.
B. On higher pitch roofs, the saw should only be used from a roof ladder.
C. If a saw cannot be safely used due to pitch, you may have to resort to using an axe.
D. Once the cutting of the vent hole is completed, the roof should be evacuated.

17. Which of the seven rules for cutting roofs below are incorrectly stated?

1. Ensure the stability of the area before entering.
2. Always have a means of escape.
3. Plan the cut before you start.
4. Arrange the sequence of cuts to keep the wind in your face.
5. Cut adjacent to the joists for maximum support and the least bounce.
6. Never step on the cut.
7. Don't cut the roof supports.

A. all except choices 1, 2, 6, and 7
B. 2, 3, and 4
C. 1, 2, 6, and 7
D. 1, 2, 3, 4, and 5

18. All but one of the following are good reasons for calling for assistance to reinforce a 10-person crew at a house fire. Which choice does not belong?

A. The fire has possession of two rooms and is blowing out the windows.
B. The fire is involving an attic or a cockloft.
C. A firefighter is seriously injured.
D. A serious fire is present in a very large home.

19. Draw a quick cut.

15 MULTIPLE DWELLINGS

Questions

1. Multiple dwellings house at least how many families?

   A. one
   B. two
   C. three
   D. four

2. When do these occupancies pose a high life hazard?

   A. daytime
   B. nighttime
   C. after midnight
   D. all hours

3. Multiple dwellings pose all of the following difficulties for firefighters except for which choice?

   A. Pipe recesses allow fast vertical fire spread.
   B. Combustible siding allows rapid fire spread to cellars.
   C. There is a time delay required to cover large buildings.
   D. Recognizing converted dwellings is difficult.

4. The term SRO (single-room occupancy) indicates all of the following to responding firefighters except for which choice?

A. the presence of padlocks on the front door
B. the need for much forcible entry
C. the potential for large losses of life
D. the need for fast reinforcements

5. All but one of the following building features promote fast spreading fires in multiple dwellings. Which choice does not belong?

A. enclosed interior stairways
B. light and air shafts
C. unprotected shafts such as dumbwaiters and compactors
D. pipe chases and channel rails

6. What are the two greatest threats to the structure in the event of fire?

1. compactors
2. dumbwaiters
3. pipe chases
4. channel rails

A. 1 and 2
B. 1 and 3
C. 2 and 3
D. 3 and 4

7. Why is the furred-out space around an I-beam more of a threat than a soil-pipe chase?

A. It is wider.
B. It goes to the cockloft.
C. It is more likely that fire will enter it.
D. It is harder to find.

8. Initial operations at multiple dwellings should focus on which choice?

A. removing all occupants
B. raising portable ladders to all persons above the fire
C. getting the first hoseline in place to protect the interior stair
D. all of the above

9. You are the first arrive at a fourth-floor fire in a six-story apartment house. The door to the fire apartment must be held closed while occupants descend the open stairway. What action should you order your ladder company members to take while waiting for the remaining civilians to clear the stair? (Note! The roof bulkhead has not been opened yet.)

A. Return to the street and raise ground ladders.
B. Return to the street and stretch another hoseline.
C. Proceed to the floor above the fire to search for trapped occupants.
D. Vent the window on the stair landing just above the fire floor.

10. What is the best method of forcing entry at a multiple dwelling?

A. through-the-lock
B. brute force
C. conventional
D. with a hydraulic forcible-entry tool

11. What is the preferred size of attack hose for the average one-apartment fire in a Class III multiple dwelling?

A. 1½ in.
B. 1¾ or 2 in.
C. 2½ in.
D. 3 in. or larger

12. As the incident commander, you arrive to find fire venting from four windows (two apartments) on the third floor of a six-story Class 3 multiple dwelling. What line should you order the first engine to stretch?

A. 1¾-in. to the fire floor
B. 1¾-in. to the floor above
C. 2½-in. to the fire floor
D. 2½-in. to the floor above

13. Stretching a hoseline to the proper location at a large multiple dwelling is critical. When attempting to stretch a hoseline to an upper floor of a large building with several staircases, what action should you take to locate the proper stair?

A. Use a bottle of rope to haul the line up.
B. Stretch the second line immediately to the floor above to stop extension.
C. Stretch via the staircase nearest the front door.
D. Have a member proceed to the second floor and locate the proper apartment line.

14. Of the three types of staircases, which is the greatest aid to firefighters?

A. isolated stairs
B. transverse stairs
C. single stair
D. wing stairs

15. A stair that maintains a constant relationship to floors and apartments on each floor is called what type stair?

A. return stair
B. scissor stair
C. isolated stair
D. wing stair

16. Which statement concerning hoseline stretching and operations in large multiple dwellings is most correct?

A. The first line should be stretched to the fire floor via any available route.
B. The second line should be stretched by the same route as the first to the floor above.
C. The third and fourth lines, if needed, should be stretched via an alternate route.
D. The fifth and sixth lines should be stretched by the same route as the first two.

17. At a serious apartment fire, the second line might correctly be stretched to any of these locations except one. Which choice does not belong?

A. the fire floor, to back up the first line
B. the fire floor, to other involved apartments
C. the fire floor, via a route opposite the first line
D. the floor above, to handle extension

18. At a serious top-floor fire, where the need for many additional handlines is evident, how should the third or fourth lines be stretched?

A. by way of the interior stair
B. from an outlet in a platform basket
C. off the tip of an aerial ladder pipe
D. by rope on the building exterior

19. Multiple dwellings are often built with different wings, arranged in various shapes. Which statement below is incorrect?

A. The worst-case scenario in these buildings occurs when fire in one wing of an H-shaped building spreads in two directions at once, because this wreaks havoc on fire operations.
B. An E-shaped building has three wings, connected by two throats.
C. Items like staircases, elevators, compactors, and incinerators are often found in the throat.
D. A throat is a good location to stop fire spread between wings.

20. For a heavy cockloft fire, an elevating platform is a must. All of the following are the advantages of such devices for this use except which choice?

A. It provides a very stable work platform.
B. It provides close observation of the master stream.
C. It permits extremely accurate operation of the stream in all directions through an opening.
D. It allows trench cuts to be made easily from the basket.

21. At all serious fires in multiple dwellings, two members should immediately proceed to the roof. What is the correct sequence of actions they should take?

1. Force the bulkhead door.
2. Vent any skylights atop the bulkhead.
3. Search the stair landing for victims trapped by a locked bulkhead.
4. Examine rear and sides for trapped occupants or extending fire.

A. 1, 2, 3, 4
B. 1, 3, 2, 4
C. 1, 2, 4, 3
D. 4, 2, 1, 3

22. What is the most preferable means of reaching the roof of a fire building located in the middle of a row of attached brick buildings?

A. aerial ladder

B. the interior stair of the fire building

C. the stairway in an adjoining building

D. fire escape

23. For a serious lower floor fire in a four-story apartment house, the roof team should take the following actions after they have vented over all vertical shafts. Which choice correctly describes the actions to take in this situation?

A. They should descend via the fire escape to begin searching the fire apartment.

B. If no fire escape is present, or if fire conditions prevent its use, they should descend the interior stairs from the bulkhead.

C. If they are unable to descend due to fire conditions, they should return to the street by the means they used to reach the roof, then make their way up the interior stair.

D. If they are unable to descend, they should immediately begin cutting the roof.

24. For a serious top-floor fire in a four-story apartment house, the roof team should take the following actions after they have vented over all vertical shafts. Which choice incorrectly describes the actions the roof team takes in this situation?

A. First, vent top floor windows from the aerial device.

B. Second, begin roof cutting.

C. If descent via fire escape is possible, venting and search from this area is preferred when fire is in proximity to the apartment entrance door.

D. After the roof hole is cut, be sure to push down the ceilings below.

25. Which of the following choices is not one of the major collapse threats at multiple dwellings?

A. vacant buildings due to rot, previous fires, etc.

B. collapse of older floor and roof systems

C. lightweight construction

D. cornice collapse

26. All but one of the following are difficulties present in Class 1 (fireproof) multiple dwellings. Which choice does not belong?

A. Early collapse of floors and roof is possible.

B. Sturdier doors and locks make forcible entry more difficult.

C. The building holds tremendous heat, radiating it back at firefighters.

D. Scissor stairs cause disorientation and hose stretching problems.

27. What is the most likely method of fire extension in Class 1 multiple dwellings?

A. Flames burn through to the floor above.

B. Flames burn through walls to the adjoining apartment.

C. Flames travel via channel rails to the cockloft.

D. Auto exposure out windows to the floor above.

28. Class 1 buildings that have a strong wind blowing in toward the fire apartment require extreme care in the horizontal ventilation. All but one of the following are good actions to take to evaluate the effect that venting the fire apartment windows will have. Which choice does not belong?

A. If outside, check the wind direction.

B. If in the fire area, vent upwind first, then downwind.

C. If on the floor above the fire and attempting to vent the fire area, position all the doors on this floor the same way the doors on the fire floor will be, open from the stair, through the apartment, to the fire area.

D. If the wind will blow in, do not vent the windows.

29. A chief officer discussing severe, wind-driven high-rise fires with several of her company officers made the following statements regarding these disastrous fires. She was least correct in which choice?

A. These fires have occurred in Class 1 residential buildings from 10 to more than 40 stories high.

B. These fires can occur on any floor of a building.

C. High wind (over 25 mph) is needed to fan these fires, which usually occur in winter.

D. Air flow through the building is the culprit, and it can create temperatures in excess of 2,500°F from standard residential fire loading.

30. Which are correct methods of recognizing sandwich apartments?

1. observing three adjacent doors in a hallway
2. observing rows of windows every three floors
3. observing fire balconies instead of fire escapes

A. all of the above
B. 1 and 2 only
C. 2 and 3 only
D. 2 only

31. Which choice below most accurately reflects the correct description and tactics for fires in duplex, triplex, and sandwich apartments?

A. If the fire apartment has the bedrooms on the lower level, occupants are more seriously exposed to fire and products of combustion.
B. If the apartment has an interior staircase, it is required to be enclosed.
C. All sandwich apartments have an open interior stair.
D. Where fire on one level of a duplex or triplex apartment blocks the only exit, it may be possible to breach a wall from an adjoining apartment to reach trapped occupants.

32. Choose the most correct statement regarding the actions you should advise the occupants of apartments to take in a Class 1 building.

A. The occupants should stay in their own apartments.
B. The occupants should immediately evacuate the building.
C. The occupants of apartments other than the fire apartment should stay in their apartment.
D. The occupants of the fire apartment should evacuate to a separate room from the fire within their own apartment.

33. What is the most correct method of controlling panic among building occupants who are hanging out of windows remote from the fire and calling for assistance?

A. As a first option, immediately on arrival, use the apparatus public address system.
B. Use the building public address system as a last resort.
C. Use of public address systems is particularly effective in apartment houses.
D. Search teams will have to account for all apartments on the fire floor and floors above.

16 GARDEN APARTMENT AND TOWNHOUSE FIRES

Questions

1. Select the most correct statement regarding fires in garden apartments and/or townhouses.

A. Garden apartments are single-family dwellings.

B. Townhouses are multiple dwellings.

C. The term *condo* can accurately be used to describe either one.

D. Fire loading and compartmentation are similar in each.

2. Garden apartments and townhouses are similar to other residential occupancies, private dwellings, and older traditional apartments in all but which of the following factors?

A. fire loading

B. compartmentation within the building

C. the need for speed of water application

D. the many turns likely to be made with the hoseline

3. Which statement about garden apartments and townhouses is most correct?

A. Garden apartments frequently have common cocklofts.

B. Townhouses frequently have common cocklofts.

C. The entrance stair to upper floors often faces the street or courtyard, making it less crucial than staircases in older multiple dwellings.

D. Some garden apartments have interior stairs and halls, and these should be treated exactly the same as those in older multiple dwellings.

4. A chief officer lecturing on garden apartment fires made the following statements:

1. They are usually one to three stories high.
2. Some may be as tall as six stories, if of Class 3 construction.
3. Living space in each apartment is usually all on one floor.
4. There is normally one interior stair in each apartment.
5. A common crawl space may be present under several units.

The chief was correct in which statements?

A. 1 and 2 only
B. 1, 2, and 3 only
C. 1, 3, and 5 only
D. 2, 3, and 4 only

5. The same chief made these statements concerning townhouses:

1. Townhouses are similar to private dwellings, but are attached on one or both sides.
2. Townhouses typically have living areas on two or more floors connected by an open interior stair.
3. Room size and fire loading are similar to commercial buildings.
4. A common crawl space is present under several units.

The chief was correct in which statements?

A. 1 and 2 only
B. 1, 2, and 3 only
C. 1, 2, and 4 only
D. all of the above

6. Select the most accurate statement concerning construction and operations at garden apartment and townhouse complexes.

A. Poor apparatus access often limits the usefulness of any master streams.
B. Placing buildings at a 90-degree angle to each other limits fire extension.
C. Brick veneer on walls contributes greatly to fire extension.
D. Wood shake shingle roofs pose a danger of conflagration.

7. **Engine companies face many difficulties at garden apartment and townhouse fires. Select the most correct statement concerning engine operations at these fires.**

   A. Hydrant spacing and flow is often deficient outside the complex.
   B. 1¾- or 2-in. hose should be used for all fires in these buildings.
   C. Several hundred feet of hose may be required to reach the apartment.
   D. Preconnected handlines should be available to reach all apartments.

8. **Large garden apartment and townhouse complexes create serious fire problems. Select the least accurate statement from the choices below.**

   A. A 2½-in. handline with a break-apart nozzle is a key tool for dealing with those apartments that are out of the reach of a preconnect.
   B. The 2½-in. line can be crucial if heavy fire is present and extending to exposures.
   C. The 2½-in. line serves to increase friction loss when making the many bends and turns inside the structure.
   D. If a nozzle team has to back out due to heavy fire, disconnect the 1¾-in. line and immediately place the stream in operation.

9. **Select the least correct statement concerning fires in garden apartments.**

   A. Compared to older apartment buildings, garden apartments use far more lightweight construction.
   B. If fire is confined to the contents of a room, there is little likelihood of collapse.
   C. If fire has begun in or extended into the structural voids, a change in tactics is required due to the likelihood of collapse.
   D. Fire attacking the structural elements requires the tactics used in other residential fires.

10. **Garden apartments closely resemble older urban row frame houses. Which tactic, so successful in row frames, is not likely to be successful at a serious modern garden apartment fire?**

   A. Stretch 1¾-in. lines into exposures.
   B. Cut a large ventilation hole over the main body of fire.
   C. Knock down the main body of fire with a 2½-in. handline or tower ladder stream.
   D. Make a stand behind a defendable barrier such as a party wall.

11. Defensive measures at serious fires in a townhouse are best reflected in which choice?

A. Mansard roofs that wrap around division walls must be opened up to expose extending fire.

B. Most fire walls in these buildings are self-supporting, independent walls that make good locations to stop fire spread.

C. Fire spread through party walls on other than the top floor is not a problem.

D. Hose streams operating near party walls can only help the situation. Apply as much water as possible.

12. Choose the most correct statement regarding fires in common cellars or crawl spaces.

A. Townhouses are noted for common cellars or crawl spaces, which are unusual in most other occupancies.

B. The cellar or basement is often the location of the common laundry, which is the source of a significant number of fires.

C. The common cellar has few of the difficulties of the common cockloft.

D. Fire in a common crawl space calls for the rapid placement of at least two 1¾-in. handlines into the space.

13. Rows of attached buildings such as garden apartments, with rapidly spreading fire in a cockloft, call for units to be highly mobile. The fire units must be able to clearly identify their location within the row of buildings. Which statement below accurately identifies where the firefighters are operating in a row of apartments?

A. A firefighter in an apartment four buildings to the left of the fire apartment says he is in exposure 4D and needs a hoseline for extension.

B. A firefighter in the building immediately to the left of the fire building says he is in exposure 2A and the cockloft is clear.

C. A fire company operating in the two buildings to the right of the fire building reports that they are in exposure 2A and need extra personnel with hooks to pull ceilings.

D. A fire unit opening ceilings in a building two buildings to the left of the fire building reports that the fire is approaching them in exposure 2A.

14. The challenge of accurately tracking the location of units operating in widely separated locations such as rows of townhouse or garden apartments is crucial to their safety. Which statement below least accurately reflects the steps to be taken in these situations?

A. Units reporting in for assignments at such situations should be physically pointed to the building they are to operate in and be told its designation, for example, exposure 2C.

B. All personnel should note any distinguishing characteristics of the building they are entering.

C. Consider spray painting the address on each building.

D. Extra caution is needed when outside streams are to begin operating into a row of buildings to avoid impacting personnel within that building.

15. The largest losses in either garden apartment or townhouse complexes occur at what stage in their development?

A. open house stage

B. occupied stage

C. open framed stage

D. closed framed stage

16. Heavy fires involving a garden apartment or townhouse in the open framed stage are best fought with which hose stream(s)?

A. a 1¾- or 2-in. line through the front door

B. two 1¾-in. lines operating side by side

C. a 2½-in. handline with a large solid tip

D. a preconnected master stream

17. Heavy fires involving a garden apartment or townhouse in the occupied stage are best fought with which hose stream(s)?

A. a 1¾- or 2-in. line through the front door

B. two 1¾-in. lines operating side by side

C. a 2½-in. handline with a large solid tip

D. a preconnected master stream

18. A captain training her unit on firefighting tactics at garden apartment fires made the following statements concerning cockloft fires. Which statement is least correct?

A. 1¾-in. lines should be stretched to the top floor of each exposure to stop extension, while other crews work to knock down the main body of fire.

B. Early cutting of a large vent hole over the fire is not safe in lightweight buildings.

C. If roof ventilation is not possible, crews are likely to be driven out of adjoining building sections.

D. The only place to make a stop under these circumstances is behind a partition.

17 STORE FIRES—TAXPAYERS AND STRIP MALLS

Questions

1. **Store fires pose all of the same dangers to firefighters as large, nonfireproof apartment houses with what exception?**

    A. common cocklofts over the entire building

    B. built with Class 3 ordinary construction

    C. high life hazard throughout the building

    D. large floor areas

2. **Select the most correct statement concerning searches of commercial buildings.**

    A. Due to the hours of operation, occupants are awake and will always flee a fire.

    B. Occupants of cellars might become trapped if they are not warned in time, as cellars are often below the main body of fire.

    C. A thorough search of the cellar is mandatory and should be conducted the same way as in residential occupancies, emphasizing the use of guide ropes.

    D. The difference between store fires and residential fire searches is that there is more time to search commercial buildings.

3. **The term *taxpayer* most accurately describes what type of structure?**

    A. a store or other commercial building

    B. a row of large, ordinary construction, residential buildings

    C. an ordinary construction building housing several commercial occupancies under one roof

    D. a row of ordinary construction, semi-attached stores with apartments above

4. Compared to residential fires, store fires kill how many firefighters per 100,000 fires?

A. the same percentage
B. four times as many
C. eight times as many
D. far fewer

5. Select the most incorrect statement regarding fires in commercial structures.

A. Fire loading in commercial structures is heavier than in manufacturing and storage occupancies.
B. Fire loading in commercial structures is heavier than in residential occupancies.
C. These higher fire loadings demand a significant increase in hose stream flows.
D. Manufacturing and storage occupancies often use more hazardous materials than retail stores do.

6. What are the primary differences between old style taxpayers and new style strip malls?

1. The amount of combustible stock present in stores is different in each type of building.
2. Strip malls are often built of noncombustible construction.
3. Strip malls are often built without cellars.
4. Strip malls lack a common cockloft.

A. all of the above
B. 1, 2, and 3 only
C. 2 and 3 only
D. 3 only

7. Which of the following is not one of the four styles of roof construction usually found on a taxpayer?

A. standard peaked roof
B. standard flat roof
C. metal deck on bar joist
D. bowstring truss

8. **A chief officer conducting a preplanning session at an older row of stores built with a standard flat roof gave the following instructions to the fire companies assembled at the scene. Which instruction does not correctly reflect a recommended tactic for these situations?**

   A. The roof team should attempt to cut an 8×8-ft vent hole over the main body of fire.

   B. If further ventilation is needed, continue cutting additional vent holes.

   C. If the vent holes do not stop fire spread, cut a trench to limit fire spread to other stores.

   D. For a trench to be effective, it must be cut from outside wall to outside wall (or to a fire wall) and be subdivided every 4 ft.

9. **A serious fire exists in a newer taxpayer built with a metal deck on bar joist roof. The roof team reports that a portion of the roof is sagging over the fire store. Which orders below most accurately portray proper tactics?**

   A. Order the roof evacuated.

   B. Order roof forces to back away 20 ft from the danger area, then resume operating.

   C. Pull interior forces back out of the involved area, then operate in exposed stores until the main body of fire is darkened down and the steel has cooled.

   D. Pull interior forces back out of the entire building until the main body of fire is darkened down and the steel has cooled.

10. **The presence of heavy fire should prompt removal of all fire forces from the roof as well as the interior of all but which of the following structures?**

    A. a bowstring truss–roofed supermarket

    B. a novelty store built with sawn joists used for floor and roof

    C. a gift shop built with lightweight trusses

    D. a fruit store with plywood I-beams supporting floor and roof

11. **Regardless of style, all taxpayers share all but one of these common problems. Which choice does not belong?**

    A. difficult forcible entry

    B. potential backdrafts

    C. heavy fire load

    D. early collapse danger

12. A newly promoted lieutenant was instructing his very senior and experienced engine company on tactics to be used at store fires. The firefighters would be wise to correct which of the officer's statements below?

A. Each engine arriving at a serious store fire should connect to a serviceable hydrant with either its soft suction or a length of 5-in. or larger hose.

B. The use of one or two lengths of large hose has advantages over a direct hydrant connection for all engine companies responding to this fire.

C. One of the most beneficial places to spot the first pumper is in line with and across the street from the fire store.

D. Fire that is blowing out the front windows calls for use of the master stream.

13. A 2½-in. handline has several advantages over smaller lines when operating at a commercial building. Which choice is not one of these advantages?

A. the volume of water delivered and the reach of the stream

B. the ability to make tight bends and turns

C. efficiency of personnel

D. the power of the stream to penetrate obstructions

14. Forcible-entry techniques that are appropriately matched to the fire situation are found in which of the following descriptions?

A. An engine officer arriving at a severe fire that had already self-vented through the show windows ordered the plate glass door to be smashed to permit hoseline access.

B. A ladder company officer arriving at an automatic alarm at a restaurant and observing no visible fire or smoke chose to smash the plate glass door rather than the show windows to gain entry for examination of the cause.

C. Observing a moderate to heavy smoke condition visible in several adjoining stores in a row, a chief officer ordered the plate glass show windows immediately vented in all of the stores.

D. A ladder company firefighter observed heavy fire spreading to stores on either side of the original fire store. She immediately went back to the apparatus to retrieve the K-tool to force the open mesh roll-up gates protecting the stores.

15. The rear of many taxpayers is extremely heavily secured. Which choice may prove to be the fastest and least damaging means of gaining entry?

A. through-the-lock

B. breaching the cement block wall

C. conventional forcible entry

D. hydraulic forcible-entry tool

16. Which of the following are warning signs of potential backdraft?

1. heavy smoke, issuing under pressure
2. highly heated windows
3. smoke puffing out and then being drawn back in
4. large amounts of visible flame

   A. all of the above
   B. 1 and 3 only
   C. 1, 2, and 3 only
   D. 1, 3, and 4 only

17. What is the proper sequence of operations at a taxpayer fire that has indications that a backdraft is likely?

1. Direct the stream onto the fire and advance slowly.
2. Vent storefront windows.
3. Vent cockloft and store from roof.
4. Place charged 2½-in. hoselines out of danger area of blast.

   A. 3, 2, 4, 1
   B. 3, 4, 2, 1
   C. 4, 2, 1, 3
   D. 4, 3, 1, 2

18. All but one of the following structures are likely candidates for steel plating of walls and roofs. Which choice does not belong?

   A. furniture warehouse
   B. jewelry store
   C. gun dealers
   D. electronics superstore

19. Which statement below is most correct concerning tactics for store fires in taxpayers or strip malls?

   A. For very heavy fires, use 2½-in. handlines in the fire store and exposed stores.
   B. A sufficient number of 10-in.-long hooks is needed to rapidly open ceilings.

C. One line should be placed inside each exposed store.

D. For heavy fire that has also entered the cockloft, an elevating platform should be used for stream application from above the roof.

20. Cockloft fires in taxpayers are extremely fast-spreading affairs that are compounded by a number of problems. Which of the following choices does not represent one of these problems?

A. multiple hung ceilings

B. potential cockloft backdraft and ceiling collapse

C. parapet wall collapse

D. ease in locating the seat of the fire

21. Upon entering a store, you encounter a heavy smoke condition at the ceiling with little or no visible fire and only a moderate heat condition. You order the following actions taken. Which of the following actions would be least correct?

A. Make a small examination hole in the ceiling just inside the entryway to each area, to see if fire is over your head.

B. Be sure to continue to poke upward until you are sure the members are hitting roof boards.

C. After you pull the hook down, examine the head of the hook for signs of fire.

D. If the hook shows evidence of fire overhead, immediately operate the line into the cockloft.

22. Parapet walls over display windows pose serious threat of collapse. Which of the following statements about this danger is least accurate?

A. Walls above show windows are carried on steel I-beams.

B. These I-beams are often connected to other beams that run from front to rear in the cockloft.

C. Any twisting of the beam at the window, or expansion of the front-to-rear beams, can topple the parapet wall.

D. The collapse zone around a parapet wall must be at least the height of the wall.

23. Besides their ability to topple parapet walls, what is another problem posed by expanding steel I-beams?

A. They can push through solid brick or cement block walls, allowing fire spread.
B. They can push through reinforced concrete walls, allowing fire spread.
C. They can push through reinforced concrete floors, causing collapse.
D. They can push through solid brick or cement block floors, causing collapse.

24. If a serious fire has been burning for a prolonged time in a store or the cockloft, the potential for parapet wall collapse exists. Which choice concerning this event is most correct?

A. The collapse zone encompasses the entire sidewalk on all frontages.
B. The parapets most at risk are the two side walls.
C. The heating of cast iron columns precipitates most of these collapses.
D. An elevating platform basket is not subject to this danger at a cockloft fire.

25. When fire is in the cockloft of a taxpayer, how many members and saws should be committed to the roof, at least initially?

A. none initially
B. two members, one saw
C. four members, one saw
D. six members, two saws

18 HIGH-RISE OFFICE BUILDINGS

Questions

1. A chief officer discussing office-building fires with his companies made the following comments. Which statement is least correct?

A. The lessons learned from past high-rise fires usually start to apply once buildings are taller than six stories.

B. Fire-resistive construction and fire protection requirements are hallmarks of high-rise office buildings.

C. Buildings as short as three or four stories high are built exactly the same as their taller cousins.

D. The strategies and tactics that work well in the taller high-rises work well in the shorter buildings also.

2. Continuing with this discussion, the chief made these additional comments. Which statement most needs correction?

A. A company may have to respond to a high-rise office-building fire even though it does not have any high-rises in its immediate area.

B. High-rise offices are far different from other types of fires.

C. High-rise residential fires pose all the same problems as office buildings.

D. Strategies for high-rise office buildings must be based on the construction and facilities of the building.

3. **All but one of the following items should be considered essential for fighting a fire in a high-rise office building. Which choice does not belong?**

   A. one-hour SCBA cylinders
   B. 2½-in. rolled or folded lengths of hose
   C. solid tip handline nozzles
   D. high-pressure fog nozzles for master streams

4. **Some of the strategies designed for use in high-rise buildings may apply to other structures as well. Which of the following situations most correctly describes a situation that might require high-rise tactics?**

   A. a wood-frame building that has limited windows
   B. a fire-resistive building that has limited windows
   C. a fire-resistive building that is out of the reach of ladders due to setback from the street
   D. a fire-resistive building that is out of the reach of ladders due to setback from the street and where all operations must be conducted from the interior

5. **Which of the following actions should the occupants of high-rise buildings be taught not to accomplish in the event of fire?**

   A. how to safely use elevators during a fire
   B. how to transmit an alarm and warn others
   C. how to obtain fresh air if they are trapped
   D. how to close vents and seal cracks in their rooms

6. **What items are included in the high-rise strategic plan?**

   1. Determine the specific fire floor.
   2. Verify the location of the fire before committing handlines.
   3. Take control of evacuation.
   4. Gain control of building systems.
   5. Confine and extinguish the fire.

   A. 1, 2, and 3 only
   B. 1 3, and 4 only
   C. 1, 3, 4, and 5 only
   D. all of the above

7. **The controlled evacuation of a high-rise building is a difficult task. Which of the following does not correctly reflect this effort?**

   A. Begin immediate evacuation of the fire floor and the floor directly above.
   B. Evacuation of the intervening floors (between the fire area and the top floors) is not necessary.
   C. Prevent panicky exit by those not endangered by the fire.
   D. Search the fire floor and all floors above the fire.

8. **Some means of preventing uncontrolled evacuations include all but which choice?**

   A. postfire education of occupants as to the actions they should take
   B. zoning of the fire alarm system so that alarm bells sound only on the fire floor and one or two floors above, plus the lobby or fire command station
   C. use of public address speakers in stairwells, elevators, and public areas
   D. teaching occupants the proper behaviors before an incident

9. **For effective operations, fire departments must be able to control a number of building systems at a high-rise fire. Which of the following is not one of the major systems to be placed under fire department control during an incident?**

   A. elevators
   B. heating, ventilating, and air conditioning (HVAC)
   C. electrical
   D. communications

10. **Which of the following is not one of the recommended categories of high-rise construction?**

   A. Class 5 lightweight
   B. Class 1 lightweight
   C. Class 1 medium weight
   D. Class 1 heavy weight

11. What high-rise office building construction feature listed below is not responsible for fire and smoke spread from floor to floor?

A. curtain wall construction

B. scissor stairs

C. HVAC systems that serve more than one floor

D. access stairs

12. Which was not one of the distinguishing features of pre–World War II high-rises?

A. less compartmentation, with no hung ceilings or blind spaces

B. generally overbuilt, with structural members encased in concrete

C. numerous exits present, usually remote from each other; often fire towers

D. absence of ducts for fire and smoke to travel in, and the presence of operable windows

13. What was the most important feature of pre–World War II high-rises in limiting fire and smoke spread?

A. low fire loading

B. the absence of central air conditioners serving more than one floor

C. floor-to-ceiling partitions of two-hour rated construction

D. concrete floor slabs on metal deck and I-beams

14. As the incident commander at a severe high-rise office-building fire that involves three floors of the building, you are forced to evaluate whether or not to evacuate all firefighting personnel from the building. What information below do you need to make an informed decision?

1. What type construction is involved?
2. Are there any trusses in the fire area?
3. What type fireproofing is used?
4. What is the civilian life hazard, and how long will it take to remove the civilians?
5. What caused the fire?

A. 1, 2, and 3 only

B. 1, 2, 3, and 4 only

C. 2, 3, and 4 only

D. all of the above

15. What effect does central air conditioning have on fire operations in high-rises?

1. Windows are generally not readily able to be opened for ventilation.
2. Smoke, heat, and fire are readily spread from floor to floor, producing conflicting reports of the fire's location and sometimes panic among the occupants.
3. The automatic smoke-removal feature simplifies fire control.

A. 1 only
B. 1 and 2 only
C. 2 and 3 only
D. all of the above

16. What is the best way to prevent smoke travel through the HVAC system?

A. fire dampers inside the ducts
B. smoke dampers in the ducts
C. smoke detectors in duct intakes that shut down the system
D. manual shutdown of the system on a report of a fire

17. The HVAC system may be useful in smoke removal in high-rises. The incident commander must know all but which of the following in order to use the HVAC system for this purpose?

A. the floor layout and the locations of all stairs
B. the location of the smoke
C. the locations of firefighters and civilians
D. whether using the system will draw fire toward firefighters or civilians

18. If you arrive at a high-rise fire and receive reports of smoke on several floors, what is the first action you should direct the building engineer to take with the HVAC system?

A. Put the system into full dump mode.
B. Put the system into nonrecirculating mode.
C. Shut the air intakes on the affected floors.
D. Shut the system down.

19. What two factors outweigh all others involved in ventilation during a high-rise fire?

A. the height and area of the building
B. the fire condition and stack effect
C. the wind and fire condition
D. the wind and the stack effect

20. A chief officer teaching a class on high-rise firefighting to a group of newly promoted fire officers made the following statements. She was correct in which choices?

1. Stack effect is a natural movement of air in a building. It becomes noticeable as a building reaches about 60 ft high.
2. Stack effect is caused by heated air rising up shafts like elevators, stairs, and mechanical shafts.
3. Stack effect increases as a building gets taller.
4. Stack effect always causes smoke to rise.

A. 1 and 3 only
B. 1, 2, and 3 only
C. 1, 3, and 4 only
D. all of the above

21. What effects does vertical ventilation have on the stack effect?

A. It will prevent mushrooming on upper floors.
B. It will draw fire and smoke away from shafts.
C. It allows smoke and heat to escape out windows on the fire floor.
D. It forces smoke and heat back into the fire area.

22. Which of the following are reasons why elevator shafts should not be used for venting a high-rise building?

1. It places an elevator out of service for fire department use.
2. The open hoistway door creates a severe fall hazard.
3. The hoistway door opening is too small to vent a serious fire effectively.

A. 1 and 2 only
B. 1 and 3 only
C. 2 and 3 only
D. all of the above

23. During a critique of a recent high-rise fire, several comments were made by various participants. Which of the following choices reflect an incorrect understanding of the effects of ventilation at such scenes?

1. Stratification occurs when there is cold smoke, such as when sprinklers have operated.
2. The cool smoke may not rise and may get stuck in the neutral zone.
3. Reverse stack effect occurs when the outside air temperature is much warmer than the inside temperature and may result in smoke moving down to floors below the fire.

A. all of the above
B. 1 and 2 only
C. 1 and 3 only
D. none of the above

24. Which choice is not one of the factors that affect the decision of whether or not to perform horizontal ventilation at a high-rise fire?

A. the effect of wind
B. stack effect, which can draw fire *into* the building instead of letting smoke and heat out
C. the status of the HVAC system
D. the effect that falling glass will have

25. What is the major danger from elevator use at a high-rise?

A. The elevator car moving in a shaft may move smoke to remote areas.
B. The open hoistway door can create a severe fall hazard.
C. The elevator will stop at the fire floor.
D. The elevator will stop between floors, trapping its occupants below the fire.

26. Which choice is not one of the reasons why elevators are often used in high-rise building fires?

A. the need to reduce the response time
B. the safety and availability of a modern firefighter's service elevator
C. the fatigue factor
D. the logistical problem of moving large numbers of people, air cylinders, hose, tools, and so forth, upstairs by foot

27. What two elements of elevator construction and design can make the use of elevators safer at a fire?

1. firefighter's service elevators
2. the presence of sky lobbies below the fire floor
3. the availability of freight elevators
4. segregated blind shaft elevators past the fire floor

A. 1 and 2 only
B. 1 and 3 only
C. 2 and 3 only
D. 2 and 4 only

28. A chief officer lecturing on high-rise tactics to a group of company officers made the following statements: "The preferred hoseline for a large fire on an upper floor of a high-rise is the 2½-in. line equipped with a 1¼-in. solid tip." Which of the following choices does not justify this chief's statement?

A. This stream provides high volume and long reach at 50-psi nozzle pressure.
B. The building pump only has to supply 65 psi at the top floor standpipe outlet.
C. This stream can provide 290 gpm at only 40-psi nozzle pressure.
D. The large, open floor plans of office floors make moving such a line a simpler task than in residential buildings.

29. A very senior captain of a busy high-rise engine company was instructing his firefighters about core construction and why is it a problem for firefighters. He was least correct in which choice below?

A. Core construction groups all of the nonrentable spaces, such as stairs, elevators, electric closets, and air shafts, in a central core.
B. This results in large open spaces for fire spread.
C. Core construction keeps all the stairs close to each other, minimizing the distance that units have to traverse to stretch a line.
D. Core construction allows fire, heat, and smoke to wrap around behind units operating out of a stairway.

30. A chief officer arriving to find an office-building fire with the entire 200×200-ft. 30th floor fully involved in fire would know that the only chance of stopping such a blaze lies in what strategy?

A. pushing an aggressive, coordinated interior attack

B. getting units onto the floor above to control extension until the fire burns itself down to a point that handlines can advance on the fire floor

C. utilizing outside streams from ground level

D. utilizing outside streams from the floor above

31. Which of the following last-ditch methods of applying a stream to a large floor of a high-rise office-building fire, where units cannot enter the fire floor due to fire conditions, is least likely to be successful?

1. using a concrete core cutter to drill holes from the floor above and lowering a Bresnan distributor
2. using a concrete cutting chain saw to cut a hole from the floor below and inserting a Bresnan distributor
3. using a floor below nozzle to direct an outside stream in through a window

A. all of the above

B. 1 only

C. 2 only

D. 3 only

32. A newly promoted chief giving her officers a lesson on high-rise fires made the following comments on scissor stairs in center core construction. Which statement is least correct?

A. Scissor stairs usually alternate from one side of the core to the other.

B. Units entering the stair on the floor below the fire will be on the opposite side of the building from the fire.

C. Units on the floor below the fire will be familiar with the floor plan on the fire floor.

D. Scissor stairs like this can cause opposing hose streams on the fire floor.

33. What is an open stairway connecting two or more floors within a single tenant's occupancy called?

A. access stair

B. egress stair

C. return stair

D. scissor stair

34. **Which of these building features is not recommended to safeguard an access stair?**

A. The stair should be located within a fire-rated enclosure.
B. Connected floors should be fully sprinklered, with a separate deluge sprinkler system protecting the access stair.
C. In all cases the floors served by access stairs should be posted in the lobby for the information of firefighters.
D. The access stair should not serve more than five floors.

35. **Which choice below would not normally be one of the incident command designations at a high-rise fire?**

A. standpipe support
B. systems unit
C. stairwell support
D. search and evacuation

36. **Which of the following should be present at the operations post?**

1. operations officer and an aide
2. two separate radios: one on the frequency being used by attack forces and the other in contact with the lobby command post
3. copies of the floor plan for the fire floor and all floors above
4. additional means of communication with the command post—either telephones or special radios

A. 1 and 2 only
B. 1, 2, and 4 only
C. 1, 2, and 3 only
D. all of the above

37. **All of the following are major functions of the operations post officer at a high-rise fire except for which choice?**

A. coordination of multiple units operating from one of several stairways in use as attack stairs
B. keeping the attack moving forward by arranging an orderly rotation of personnel
C. implementing the strategy that the IC has developed
D. keeping the command post informed of the tactics being used and their results

38. All of the following are some requirements of a forward staging area in a high-rise fire except for which choice?

A. at least two attack teams and two support teams, plus an EMS sector

B. sufficient room for all these personnel

C. spare hose, tools, air bottles, etc.

D. staging area that is remote from the attack stair, with good communications with the operations post and the command post on several channels

39. All of the following are major concerns when deciding where to establish a search and evacuation post except for which choice?

A. the ability to get past the fire with a reasonable degree of safety and be able to retreat if necessary

B. the ability to communicate easily with the operations post and command post

C. the ability to communicate easily with all the personnel who are operating as part of the search and evacuation effort above the fire

D. the ability to communicate easily with units operating on the fire floor

40. Which choice is not one of the responsibilities of the search and evacuation officer?

A. Direct and control the activity of all forces operating above the fire floor and the floor immediately above.

B. Verify and record the results of searches on all floors above the fire floor.

C. Maintain contact with the operations post so all personnel above the fire can be withdrawn to safety if the fire cannot be controlled.

D. Direct units to search for any reports of people unaccounted for above the fire area.

19 CONSTRUCTION, DEMOLITION, AND RENOVATION

Questions

1. **Buildings under construction, renovation, or demolition pose unique challenges to firefighters for all but which of the following reasons:**

    A. the fire load

    B. structural problems

    C. firefighting operations

    D. the size of the structure

2. **All of the following are reasons why fires in buildings under construction, renovation, and demolition account for a large number of serious fires as well as firefighter injuries, except for one statement. Which choice does not belong?**

    A. large amounts of concealed combustibles

    B. numerous sources of ignition

    C. lack of adequate fire protection features

    D. unlimited air supply if windows are not intact

3. **Considering the potential threat of these buildings, which have work in progress, which choice explains why serious fires do not occur more often in these buildings?**

    A. Most sources of ignition are only present when workers are gone for the day.

    B. Fire watches are posted around the entire site.

    C. Fires are spotted and extinguished in their incipient stage.

    D. Power to hazardous activities may be removed during work hours.

4. Preplanning of large construction sites is vital to fire suppression efforts. What are the key items to document when visiting a structure?

1. large, undivided floor areas.
2. unenclosed vertical openings
3. limited access to upper floors
4. storage of flammable gases in temporarily enclosed areas
5. fire protection equipment not keeping pace with construction
6. buildings partially occupied

A. 1, 2, 3, and 6 only
B. 2, 3, and 5 only
C. 2, 3, 4, 5, and 6 only
D. all of the above

5. During severe fires, elevating platform streams can be extremely helpful if the fire is within their reach. What are some precautions that should be taken in their use?

1. All personnel should be removed from all points where fire could be driven at them.
2. All personnel should be removed from areas such as the streets below where the master streams could propel large pieces of debris.
3. Streams should be directed against scaffolding or shoring at close ranges in order to maximize effectiveness.

A. 1 and 2 only
B. 1 and 3 only
C. 2 and 3 only
D. all of the above

6. It may be necessary to vary the application of master streams in buildings under construction or demolition from the pattern normally followed in ordinary buildings. Select the correct statement about this operation.

A. Start the stream on lower floors and work up.
B. Start the stream on upper floors and work down.
C. Water damage is more of a concern in these buildings.
D. Use wide fog patterns to maximize steam production.

7. Using hoists or temporary elevators at construction sites can cause serious problems. How do you tell a material hoist from a personnel elevator?

A. Personnel elevators have guard rails.

B. Personnel elevators have controls in the car and carry a passenger label.

C. Material hoists are automatic.

D. Material hoists are larger.

8. A serious threat to firefighters at buildings undergoing renovation or demolition is asbestos abatement operations. Select the least correct statement concerning this process and the dangers it poses to firefighters.

A. Asbestos abatement must be done before structural components are removed.

B. Asbestos abatement exposes the firefighters and the public to a hazardous dust.

C. The operation is often done in an area that is constructed of plywood lined with plastic.

D. Entry into this space often must be made through an intermediate chamber, and limited exits add to the danger to firefighters.

9. Which is not one of the common difficulties found in the use of standpipe systems in buildings under construction or demolition?

A. Siamese connections are blocked or hidden.

B. The system is dry, and all outlet valves are often open.

C. Top of the riser is uncapped, letting water flow right out the open top, instead of through hoselines.

D. Sectional valves are open below the fire, letting water flow right past them.

10. Which of the following is one of the dangers of relying on membrane fire protection systems such as fire-rated hung ceilings in buildings under construction or renovation?

A. The ceiling panels are among the first items to be installed during construction.

B. The ceiling panels are among the last items taken out for renovations.

C. The absence of these panels allows fire to attack the structure above.

D. The use of membrane fire protection is superior to other forms, such as encasement in concrete.

11. What is one problem with using spray-on fireproofing on exposed steel structural elements?

A. Spray-on fireproofing is often scraped off to improve its function.
B. Spray-on fireproofing is rather easily removed.
C. Spray-on fireproofing clings tenaciously to steel.
D. Steel does not need spray-on fireproofing to resist a structural fire.

12. Fire is a threat to all but which type of steel found in structures?

A. steel cables in post-tensioned concrete after packing
B. steel I-beams and columns in high-rise buildings after bolting
C. steel reinforcing rods in cast-in-place concrete before curing
D. temporary steel cables bracing structural elements

13. What is perhaps the most serious collapse danger in buildings under construction?

A. collapse of a post-tensioned concrete building after packing due to fire on that floor
B. collapse of the top floor of a steel-framed high-rise in the early stages of erection due to a fire on that floor
C. collapse of just poured floor of a poured-in-place concrete building due to fire on the floor below
D. collapse of a cured reinforced concrete floor due to fire three floors below

14. For how long after reinforced concrete is poured does the danger of collapse due to fire exist?

A. 24 hours
B. 48 hours
C. 28 days
D. 48 days

15. Select the most correct statement below concerning concrete construction.

A. Poured concrete that is properly shored poses little danger in the event of a fire during the first 48 hours after pouring.
B. After 48 hours of curing, all shores may be removed from beneath the concrete.
C. It takes approximately 14 days for concrete to fully cure.
D. After concrete has fully cured, there is little danger of fire causing total collapse.

16. The practice of allowing tenants to occupy an area where there is work in progress below that floor creates severe dangers. What is one of the dangers of this practice?

A. Basic exposed pipe sprinkler systems are installed on all floors up to and below any occupied floors.

B. The lower floors are storage areas for materials being used to finish other areas.

C. Stairwell doors and doors between occupancies are rarely blocked open to facilitate movement of materials from one area to another.

D. The presence of metal construction shanties provides quite a bit of fuel to expose other floors.

17. Fires in vacant buildings kill firefighters at a rate that is up to five times greater than at occupied residential buildings. Select the answer below that lists the most correct description of why this is true.

1. Vacant buildings are often old and damaged by weather and vandalism.
2. Previous fires may have weakened the structure.
3. Structural supports may have been removed.
4. Rapid fire spread is aided by enclosing walls and doors.

A. 1 and 2 only

B. 1, 2, and 3 only

C. 1, 2, and 4 only

D. all of the above

18. Select the most accurate statement regarding fire operations in buildings undergoing demolition.

A. Just because it looks like a building doesn't mean it is a building.

B. Particularly in the case of buildings undergoing renovation, the building is nothing more than a pile of rubbish.

C. Buildings undergoing renovation never have a life hazard.

D. Buildings that have been completely gutted pose few threats to firefighters.

19. **In the author's recommended system of marking vacant buildings, the symbol below indicates to arriving firefighters that which choice is most correct action to take?**

    RO

    A. Rear open; building is of normal stability.
    B. Rear open; building has severe hazards.
    C. Roof open; building requires extra caution.
    D. Roof open; building has severe hazards.

20. **When discussing vacant building fires, a lieutenant instructing his members made the following statements. The officer should be corrected in which choice below?**

    A. Vacant building marking systems can indicate the hazards present and what type of attack should be conducted.
    B. The fact that a building is marked indicates it is vacant and operations should be conducted more slowly than at a building with a life hazard.
    C. Since buildings can be examined in daylight, before a fire, hazards are much more obvious than during a late night fire.
    D. Firefighters must rely on these inspections and on marking systems for their own safety.

20 INCINERATORS, OIL BURNERS, AND GAS LEAKS

Questions

1. Natural gas is composed primarily of which of the following?

   A. ethane
   B. methane
   C. propane
   D. nitrogen

2. When attempting to determine whether natural gases are present, a sample can be analyzed by a laboratory. The presence of what two gases would indicate that the area contains natural gas?

   A. methane and ethane
   B. methane and propane
   C. ethane and propane
   D. ethane and nitrogen

3. Which statement about natural gas is incorrect?

   A. It is colorless.
   B. It is odorless.
   C. It is lighter than air.
   D. It is toxic.

4. Natural gas has a mercaptan compound added to it to make it recognizable. What statement below is most correct in the event of a large above-ground leak?

A. The odorant can escape, making the natural gas undetectable at the leak source.

B. The odorant can create a severe fire hazard at the leak.

C. The odorant can settle out upwind, making people believe they are in danger when in fact no gas is present.

D. The odorant is added in tiny quantities, only ¼ lb per million cubic ft of gas, making it possible not to detect the leak by smell before it reaches the flammable range.

5. What problem might this mercaptan additive pose in the event of an underground leak?

A. The odorant may be filtered out by the soil.

B. The gas can migrate only short distances into buildings, manholes, etc.

C. This cannot result in a flammable mixture with no gas odor.

D. all of the above

6. Which type of gas leak poses the greatest danger?

A. gas leak with fire

B. gas leak outdoors

C. gas leak inside a structure

D. none of the above

7. What factors make large transcontinental gas pipelines burn longer than distribution main piping?

1. extremely high pressure (350–850 psi)
2. large diameter pipelines
3. long distance between valves
4. leaky valves

A. all of the above

B. 1 and 2 only

C. 1, 2, and 3 only

D. 2 and 3 only

8. Low-pressure gas systems operate at what pressure?

A. ¼ psi
B. 1 psi
C. 60 psi
D. 99 psi

9. High-pressure gas systems use regulators to control the pressure on the appliance side of the regulator. If this device fails, which choice is not one of the potential problems inside the affected structure?

A. The excessive gas pressure can cause the pilot lights to increase greatly in size, igniting nearby combustibles.
B. The excessive gas pressure can cause the pilot or burner flames to be blown out, allowing a buildup of unignited gas.
C. The excessive gas pressure can cause the burner flames to increase greatly in size, igniting nearby combustibles.
D. The excessive gas pressure can cause the pilot lights or the burner flames to increase greatly in size, allowing a buildup of unignited gas.

10. Failure of a gas regulator can be recognized by the gas odor and a hissing sound coming from the vent. What actions should the fire department take in this case?

A. Notify the utility to stop the flow of gas via remote control valve.
B. Search premises for fire, gas, and overcome victims, and vent as required.
C. Pull the electric meter to remove sources of ignition.
D. If overcome victims are found while searching, immediately radio for assistance.

11. What danger does manufactured gas pose that natural gas does not?

A. Manufactured gas is flammable.
B. Manufactured gas is odorless.
C. Manufactured gas contains carbon monoxide.
D. Manufactured gas displaces oxygen.

12. What should be the fire department's first action upon receiving a report of a gas leak?

A. Stretch a hoseline to the front door.
B. Vent the affected area.
C. Stop the source of the gas.
D. Notify the utility.

13. In the event that a gas leak is found in a stove, what is the preferred point of control?

A. appliance cock
B. meter cock
C. curb cock
D. street valve

14. An odorant is added to natural gas that allows us to smell it at about the time it reaches what percentage of the lower explosive limit?

A. 1%
B. 4%
C. 25%
D. 100%

15. All of the following are potential sources of ignition for a flammable gas–air mixture except which choice?

A. older fire department portable radio (handie talkies) when transmitting
B. fire department pagers being activated
C. turning off light switches
D. static electric spark from walking across a rug

16. Under what circumstances might it be prudent to not vent immediately where gas is leaking into a structure?

A. when the gas–air mixture is already well below the lower explosive limits (too lean)
B. when the gas–air mixture is already well above the upper explosive limits (too rich)
C. when the gas leak is too small to be controlled easily
D. when the sources of ignition are already removed

17. The first engine to arrive at a serious indoor gas leak must be prepared to supply attack lines if an explosion occurs. Which is not a factor affecting their placement and use?

A. Hoselines should be placed where they will immediately be able to enter the building.
B. The lines must be long enough to cover the entire building.
C. The water supply must be able to provide needed flows.
D. Apparatus should be positioned to protect members from blast.

18. At a recent multiple-alarm fire involving a very large, high-pressure natural gas transmission main, the fire department took the following actions. They were correct in all but which choice?

A. evacuated the danger area
B. used hose streams (solid tips at long distance) to protect exposed structures
C. examined nearby structures for evidence of gas leaking into the building
D. extinguished the fire only by closing a valve they located in a nearby street

19. A leak on a plastic gas pipe is especially dangerous due to what phenomenon?

A. odorant filtration
B. high-pressure cracking
C. freezing and thawing
D. static electric buildup

20. At an excavation site a backhoe operator punctured a high-pressure steel gas pipe, causing a leak, which ignited. All of the following fire department actions would be correct at this scene except for which choice?

A. Fire department members used dry chemical to extinguish the fire while rescuing the trapped backhoe operator.
B. Fire department members used CO_2 to extinguish the blaze in order to allow utility workers to shut a street valve.
C. The fire department stood by until the gas company shut down remote valves, stopping the flow of gas.
D. Fire department members under water curtain protection attempted to plug the leak.

21. One gallon of LPG (liquefied petroleum gas) when completely vaporized will form how many gallons of pure vapor?

    A. 27 gallons
    B. 270 gallons
    C. 1,700 gallons
    D. 2,700 gallons

22. As the temperature of propane liquid inside an LPG cylinder rises from 70° to 100°, what most nearly happens to the pressure exerted on the cylinder?

    A. The pressure rises 30 psi.
    B. The pressure rises 30%.
    C. The pressure doubles.
    D. The pressure quadruples.

23. What is the only way to prevent a BLEVE (boiling-liquid, expanding-vapor explosion) of a fire-exposed LPG cylinder?

    A. Cool the lower liquid portion of the cylinder.
    B. Extinguish the fire.
    C. Activate the relief valve on the container.
    D. Remove the heat from the upper vapor space of the tank shell.

24. An engine company approaching an LPG barbecue fire in a backyard of a private home should be prepared to continuously supply at least how many and what size lines for the duration of the incident?

    A. one 1¾-in.
    B. two 1¾-in.
    C. one 2½-in.
    D. three 1¾-in.

25. In the event of a propane leak from a cylinder without fire, hose streams may be needed for which purposes?

A. to direct vapors toward sources of ignition
B. to dilute the vapor–air mixture with air that was entrained in the fog streams
C. to cool the cylinder
D. to protect the hoseline crew from flying shrapnel

26. What is the most destructive type of event associated with a propane leak or fire?

A. a fire
B. a BLEVE
C. a vapor–air explosion
D. a MOAB

27. How do residential oil burners get fuel oil to burn very readily, even though it is well below its flash point?

A. by preheating the oil with unused steam
B. by preheating the firebox
C. by atomizing the oil with a high-pressure gun
D. by using a fan to blow large quantities of air into the flame

28. When an emergency develops with an oil burner, how should the fire department stop its operation?

A. through use of the emergency or remote control
B. choice A plus closing fuel valve at the tank
C. choices A and B plus inserting a nonconductor between the high-voltage electrodes
D. all of the above

29. Units responding to a smoky cellar found a small fire burning inside the oil burner as the source of the smoke. How should they deal with this minor blaze?

A. Apply water fog to the burning oil.
B. Apply foam to the burning oil.
C. Apply dry chemical to the burning oil.
D. none of the above

30. A captain training a newly assigned firefighter made the following statements regarding compactors and incinerators. The captain was correct in all but which choice?

A. Incinerators are vertical shafts that allow rubbish to be dropped to the lowest level where it is burned.

B. Compactors resemble incinerators in that they are both shafts that rubbish can be dropped into, through doors located in or near the hallway.

C. Compactors often have an auxiliary burner at the base to make the rubbish burn more cleanly.

D. A blocked chute with fire in a shaft can cause smoke or fire to exit the shaft doors and fill the halls, creating life-threatening conditions.

31. Select the least correct statement concerning carbon monoxide (CO) responses below.

A. Carbon monoxide is a colorless, odorless gas.

B. Carbon monoxide is tasteless.

C. Carbon monoxide is non-life-threatening.

D. Carbon monoxide is caused by fuel-burning appliances.

32. In the absence of a specific department protocol, a ladder company responding to a reported activated carbon monoxide detector in a home would be operating properly if they performed which of the following actions?

A. questioned the occupants to determine if anyone was feeling ill

B. sent a member with a CO detector to examine all parts of the structure

C. had the member with the detector start the investigation at the heating plant

D. finding less than 35 PPM of CO, allowed the occupants to return after venting the area

33. Proper procedures at creosote-fed chimney fires include all of the following except which choice?

A. Drop a plastic bag of Class ABC dry chemical powder down the flue from above.

B. Shoot a stream of Class ABC powder up the flue from below.

C. Use a fog nozzle up the flue from below.

D. Lower a special chimney nozzle down the flue from above.

21 ELECTRICAL FIRES AND EMERGENCIES

Questions

1. **Firefighters must have a basic understanding of electricity in order to perform safely on the scene of any structural fire and many emergencies. Select the choice below that most incorrectly describes electricity or its terminology.**

    A. Electricity has electromotive force or pressure, measured in voltage.
    B. Electrical flow rate is measured in amperes.
    C. Electrical resistance to flow is measured in ground.
    D. All three terms in choices A, B, and C are related and affect each other.

2. **When discussing the relationships among voltage, amperage, and resistance, a captain who is also an electrician made the following statements. Which statement was least accurate?**

    A. Amperage is determined by the amount of voltage the conductor is delivering and the resistance the wire is offering.
    B. A large wire can conduct a large amount of current.
    C. A wire with high voltage behind it and very high resistance will have very high amperage.
    D. A wire with low voltage and high resistance will have low amperage.

3. Select the incorrect statement below concerning electricity and its conductors.

A. You can tell how much current a wire carries by looking at how large the wire is.

B. Voltage can vary greatly in the same size wire.

C. The danger posed by electricity depends on variables such as moisture on the body, or whether you are standing in a puddle or on dry wood.

D. Anything less than 600 volts is considered low voltage by many utilities.

4. Select the most correct statement regarding the electrical system.

A. An open circuit does not permit electricity to flow due to an open switch or break in a wire.

B. A short circuit does not allow electricity to flow due to a break in a wire.

C. The utility ground is a wire installed by the utility to conduct electricity to the shortest path.

D. A house ground is a condition where the entire house is electrically grounded to the utility system.

5. When discussing elements of the electric utility's system, a lieutenant made the following statements. She was correct in which choice(s)?

1. Wires leaving generating stations can be at voltages over 300,000 volts.
2. These wires are not insulated.
3. The voltage is reduced at substations by transformers.
4. In the substations, insulated buss bars conduct electricity around the site.

A. 1 only

B. 1 and 2 only

C. 1, 2, and 3 only

D. all of the above

6. Which statement below does not accurately describe elements of an electrical distribution network?

A. An overhead utility pole can carry a variety of types of wires at many different voltages.

B. Typically, the higher the wire is located on a pole, the higher the voltage it carries.

C. The number and size of porcelain insulators on a wire are indications of its voltage.

D. The size of the wire is an indication of its voltage.

7. **The human body does not stand up well to close contact with electrical current. Which statement below does not correctly illustrate a hazard of electricity to humans?**

   A. Electrocution interferes with the body's own currents that cause the heart to contract.

   B. Electrocution can also cause expansion of muscles, causing the body to freeze in contact with a wire.

   C. Electricity can also cause severe burns from contact with high voltages or from arcing.

   D. Electricity can also cause explosive amputation of a limb if current enters one point on a limb and exits at another.

8. **A captain of a very busy ladder company was conducting a class for the members of the unit on electrical hazards in firefighting and made the following comments regarding possible injurious effects from electrical contact. Which statement is least accurate?**

   A. Electrocution is affected by several variables, including the amount of current passing through the body, the condition of the skin, and the duration of the contact.

   B. The path the current takes through the body is an important factor in determining the survival of a person contacting electricity.

   C. Current flow that traverses the chest is severe since it affects the heart and other vital organs.

   D. Current that enters through the head and exits through the feet damages the brain and nervous system; current flow the opposite direction is not a problem.

9. **Not to be outdone by the ladder company captain, the lieutenant of the engine company chimed in with the following comments about electrical arcing. He was correct in which choice?**

   A. Electrical arcs occur while current is following its normal path.

   B. Electric arcs such as from high-voltage wires can be very dangerous, but low-voltage sources such as from a car battery pose no danger.

   C. Static electricity discharges are an example of a low-voltage arc that is completely harmless, if irritating.

   D. An arc can occur between two wires, or travel from a wire to ground, or can cross an air gap to reach a path to ground.

10. Which statement below would be most correct regarding the phenomenon known as voltage gradient?

A. Voltage gradient results from current traveling through the ground from a downed high-voltage wire or other source.

B. Voltage radiates outward from the source in concentric circles.

C. The closer a person is to the source, the lower the voltage sensed.

D. If you detect a tingling sensation, immediately stop all movement and walk diagonally away from the scene.

11. Operational procedures at electrical fires and emergencies involving substations, transformers, manholes, and utility poles are incorrectly stated in all but which of the following comments?

A. The safety of the operating personnel is paramount.

B. The tactical position is generally offensive until power is shut down.

C. The need for defensive tactics is rare and is only justified if no life hazard exists.

D. At these incidents most of the damage is done after our arrival, and our concern is the protection of exposures.

12. While conducting a critique of a fire in a large electric power generating station, a chief officer made the following statements. Which statement(s) are correct regarding the hazards that may be present therein?

1. Some of the oldest transformers might be contaminated with PVC.
2. Steam is often present at 212°F.
3. High-pressure steam hampers both visibility and hearing if it escapes.
4. Many flammable materials may be present including helium gas, which burns with such a pale flame that it can be nearly invisible in daylight.

A. all of the above

B. 1, 2, and 3 only

C. 2 and 3 only

D. 3 only

13. Which of the following most incorrectly describes proper preparations and procedures for a fire at an electrical generating station or substation?

A. Fire departments should try to meet any utility representatives at a predetermined location inside the generating station.

B. The watch supervisor or other senior utility representative who knows the plant should brief firefighters on the situation and tell them how to get to the incident site.

C. At unstaffed stations, wait outside the site until knowledgeable utility personnel arrive. There is no life hazard at this incident.

D. No firefighter should ever enter a power plant or substation without being escorted by a knowledgeable utility representative.

14. Choose the least correct action to take at fires in a generating station or substation.

A. The firefighters wait outside the facility for a knowledgeable power company representative to brief them and escort them to the incident site.

B. The firefighters leave all metal tools outside the plant.

C. The firefighters ensure that only wooden ladders are used to reach a small fire atop a metal superstructure.

D. The firefighters ensure that all hoselines stretched are equipped with a fog tip.

15. Transformers of various sizes are used throughout the electrical distribution network, and since they create heat, they are sometimes involved in fires. Select the least accurate statement regarding transformer fires from the choices below.

A. Transformers are often filled with a nonconducting fluid, sometimes oil based.

B. Large transformers can contain 20,000 gallons of oil, which, if it ignites, is easily contained with fog streams.

C. An oil leak or fire can be a serious hazardous materials incident if the oil contains PCBs (polychlorinated biphenyls).

D. SCBA use is mandatory at any transformer fire, even outdoors, and personnel and equipment may need to be decontaminated after use.

16. A newly promoted lieutenant, wishing to impress her new unit, gave a drill regarding a recent transformer fire. She made the following statements. Which choice is least correct?

A. Assume a defensive position, with the apparatus upwind and uphill from the transformer if possible.

B. Secure a water supply, but do not flow any water until assured by utility personnel that the power is off.

C. Once power has been removed, use dry chemical or foam streams to extinguish the burning oil.

D. Do not begin overhaul until advised by the utility about the presence of PCBs; consider using hazardous materials personnel for overhauling.

17. A chief officer in a northern city, preparing his units for the upcoming winter, made the following statements about manhole fires. Which statement needs correction?

A. Manhole fires are very high in carbon monoxide (CO) production and produce other toxic gases from the PVC wire insulation.

B. The toxic gases can produce hydrochloric acid in the lungs, so SCBA use is mandatory.

C. CO can follow conduits into buildings, be sure to monitor for CO in cellars. Upper floors are not affected.

D. CO is an extremely flammable gas that can explode with great force.

18. One situation involving electricity that may result in a structural fire is an open ground in the service line, which causes the electrical current to seek another path to ground through other objects like pipes and metal objects. Choose the least accurate statement below concerning operations at an open ground.

A. Units encountering sparks jumping between a steam pipe and a metal ceiling might expect an open ground. They should realize that there may be similar situations occurring at other locations in the building simultaneously.

B. To alleviate the danger, firefighters should immediately open the building's main circuit breaker.

C. Handlines should be stretched to cover the entire building, as fires can break out in several locations at once.

D. Be sure to examine the surrounding buildings' electrical services, and also look for dimming of light fixtures and overheating of the electrical services.

19. An engine company arrived first-due at an automobile versus pole accident that caused the electrical wires on the pole to break and fall on the car. The acting officer in the right front seat gave the following commands. Which one could have caused a tragic outcome if complied with as stated?

A. positioned the apparatus to detour traffic while requesting law enforcement assistance for crowd and traffic control

B. created a safety zone around the downed wires that was at least 10 ft in all directions from the wire

C. ordered a handline with a fog nozzle stretched but kept at least 25 ft from any electrical equipment involved

D. directed the occupants of the car to remain inside and clear of any metal parts until the power could be removed

20. A captain investigating what smelled like an overheated ballast gave the following orders. She was not correct in which choice?

1. ordered the thermal imaging camera used to check the incandescent fixtures for heat
2. began by looking at fixtures with all bulbs in place, but emitting only a faint glow
3. focused on a fixture with a dark-colored oil stain on the diffuser

A. all of the above

B. 1 only

C. 2 only

D. none of the above

22 STRUCTURAL COLLAPSE

Questions

1. **Structural collapse is one of the most feared occurrences on the fireground, resulting in a large percentage of multiple casualty incidents. Which statement below regarding collapse is least accurate?**

   A. Past collapses were extremely devastating; we are not likely to experience such tragedies again.

   B. Statement A, plus the fact that changes have been made in construction that improve the fire resistance of buildings.

   C. Firefighters have a responsibility to learn the causes and warning signs of collapse.

   D. Another component of understanding collapse is to learn the steps to take to rescue anyone caught in the debris.

2. **Regarding structural collapse of high-rise office buildings, which statement is most accurate?**

   A. Buildings of Class 1 construction show the greatest resistance to collapse.

   B. Modern Class 1 construction allows enough time for a long stair climb, evacuation, fire control efforts, and a long climb down before collapse.

   C. Buildings that were constructed with a steel-reinforced, poured-concrete skeleton are most resistant to collapse from fire at all stages of construction and occupancy.

   D. The exception to choice C is precast concrete construction after occupancy.

3. Rank the five classes of construction in descending order of resistance to collapse.

1. Class 1: fireproof
2. Class 2: noncombustible
3. Class 3: ordinary construction (brick and wood joist)
4. Class 4: heavy timber
5. Class 5: wood frame

A. 1, 2, 3, 4, and 5
B. 1, 3, 4, 5, and 2
C. 1, 4, 3, 5, and 2
D. 1, 4, 3, 2, and 5

4. Generally, steel-reinforced, poured-concrete buildings are superior in their resistance to collapse. What is an accurate exception to this statement?

A. Fire involving the plywood forms of a just poured floor could precipitate spalling.
B. Fires involving an occupied concrete building pose a severe threat of collapse.
C. Fires involving a floor shored with wood posts every 8 ft are likely to be severe due to the extreme fire loading.
D. Fires involving the formwork of the most recently poured floor could cause total pancake collapse.

5. Fires in heavy timber structures generally do not endanger interior fire forces with structural collapse. What is a correct major exception to this statement?

A. Fires in heavy timber buildings cause collapse as the fire becomes so severe that it drives the firefighters out of the building.
B. A heavy timber building that has had multiple serious fires could collapse from a later, less serious blaze.
C. The size of the load-bearing elements in heavy timber buildings leaves them prone to failure from a localized fire.
D. Collapses in these buildings are usually localized; the bearing elements transmit their loads well to lower floors and bearing walls.

6. Which choice accurately depicts the behavior of steel when considering the collapse potential due to fire exposure?

1. Steel expands when heated, which can push walls or columns over.
2. When heated to over 1,000°, steel sags and twists, dropping its load.
3. When cooled, steel contracts to its original length, possibly pushing a beam off its support if it was severely sagging just prior to cooling.

A. 1 only
B. 1 and 2 only
C. 2 and 3 only
D. all of the above

7. Which choice below correctly describes a framed structure?

A. a structure built nearly entirely of wood, with wooden bearing elements
B. a structure built largely of combustible materials supported by bearing walls
C. a structure where most of the weight is carried on a skeleton of wood
D. a structure where most of the weight is carried on a skeleton or framework of steel or concrete

8. In which type building, framed or unframed, is a collapse likely to be more serious?

A. an unframed building, because it is built of all combustible materials
B. a framed building, because it is built of all combustible materials
C. an unframed building, because it depends on bearing walls
D. a framed building, because it depends on bearing walls

9. Place the following structural elements in their proper order ranging from most to least critical in terms of structural collapse danger.

    1. columns
    2. bearing walls
    3. beams and joists
    4. girders
    5. floor and roof decks

    A. 1, 2, 3, 4, 5
    B. 1, 2, 4, 3, 5
    C. 2, 1, 4, 5, 3
    D. 2, 1, 4, 3, 5

10. What is a serious drawback of cast iron as a structural element?

    A. Cast iron has great compressive strength.
    B. Cast iron is easily cast in a mold.
    C. Cast iron is a very brittle material.
    D. Cast iron can be bolted together in prefabricated sections.

11. All of the following are major causes of collapse at fires except which choice?

    A. fire damage to wood structural members
    B. heating of protected steel
    C. expansion of absorbent stock and overloading of floors
    D. vibration and impact load

12. All of the following are potential warning signs that collapse may occur except which choice?

    A. hazardous occupancies
    B. truss or lightweight construction
    C. heavy fire present for over 20 minutes in Class 1 construction
    D. plaster sliding off walls, windows cracking, and doors swinging open or closed

13. A door that suddenly swings open or closed on its own, or a window that suddenly cracks without being struck, is a sign that a building is moving. This type of warning is most likely to occur in which type building?

A. a haunted house

B. a wood-frame home

C. a high-rise office building

D. a Class 2 taxpayer

14. After the order to evacuate a structure has been given, the incident commander must take all but which steps to protect the members, in the event a structural collapse is imminent?

A. The incident commander must order an emergency evacuation in time for all members to safely withdraw from the scene with all their tools and equipment.

B. The incident commander must contact each unit operating on the fireground and confirm their receipt of the evacuation order.

C. If a unit fails to acknowledge, its members must be sought out and advised.

D. The incident commander must ensure that all members observe safe exterior collapse zones.

15. Select the least correct statement regarding emergency evacuation signaling.

A. One method is to order all apparatus on scene to turn on all their visual warning signals for 30 seconds.

B. Members who have withdrawn from the building should rejoin their units and prepare for a head count.

C. Emergency evacuation signaling could prevent firefighters from hearing critical messages.

D. Emergency evacuation signaling is needed because not everyone has a radio, and even those who do may not hear a message due to background noise.

16. How large should a collapse safety zone be?

A. at least as large as the height of the facing wall

B. at least 1½ times the height of the facing wall

C. at least 3 times the height of the facing wall

D. at least 1½ times the height of the facing wall for its full length

17. Which members must be particularly closely monitored to prevent them from encroaching on a collapse safety zone?

A. rescue company personnel

B. engine company personnel operating in exposures

C. ladder company personnel operating on the roof of an exposure

D. ladder personnel operating in the basket of an elevating platform

18. As a minimum, how far away should ground forces be from a wall of a bowstring truss roof that has hip rafters resting on it?

A. at least the height of the wall

B. at least 1½ times the height of the wall

C. at least 2½ times the height of the wall

D. at least 4 times the height of the wall

19. Where separation distances between buildings do not allow personnel to be placed out of the collapse zone, other options are available for applying hose streams. Select the most correct choice below for this situation.

A. Use master streams from outside the collapse zone and cellar nozzles or distributors from above.

B. Use streams from flanking positions or positions above the height of the wall, but beware of the fireball, especially downwind.

C. Use hoselines or master streams from adjacent buildings.

D. Use hoselines, master streams, or distributors from flanking positions.

20. Of the major types of collapse, in which type is it often simpler to locate a victim?

A. lean-to collapse

B. V-shaped collapse

C. pancake collapse

D. individual collapse

21. In the event of an explosion that has pushed out one bearing wall, which type of collapse are you most likely to find?

A. V-shaped collapse
B. pancake collapse
C. lean-to collapse
D. individual collapse

22. A collapse that results from beams burning through in the middle or from an overloaded floor would probably produce which type of collapse?

A. V-shaped collapse
B. pancake collapse
C. lean-to collapse
D. individual collapse

23. Where are you most likely to find survivors in a lean-to collapse?

A. in individual voids created by light furniture and other debris
B. on the floor below the collapsed one, against the remaining standing wall
C. on the collapsed floor, against the remaining standing wall
D. on the floor below the collapsed one, along the center axis

24. Which choice below is not one of the stages of the collapse rescue plan?

A. surface victim removal/accounting
B. search of debris pile
C. selected debris removal/tunneling
D. general debris removal

25. Before initiating any attempt to rescue victims in a collapse, a size-up of the scene must be made. This is necessary to protect the victims as well as the rescuers, and to select the best approach to the victims. Which of following is not one of the items that must be considered as part of this size-up?

A. What happened and where?
B. Who and how many are missing?
C. Where were they last seen, and can they still be alive?
D. Are there problem occupancies with finances involved?

26. A victim tracking coordinator is essential at a collapse. This person should perform all of the following tasks except which choice?

A. Interview all persons being removed from the scene.

B. For choice A, note the name of each person, identifying features, injuries, where transported, and method.

C. Ask about who was nearby when the collapse occurred, how the victim escaped, and if he or she can direct searchers to this area via the fastest route.

D. By keeping track of live victims leaving the area, the search can be halted as soon as all live victims are accounted for, thus ending any threat to searchers' lives.

27. Which choice below incorrectly describes why selected debris removal, trenching, and tunneling are performed after all accessible voids have been searched?

A. Void searches are much faster since no digging is required.

B. Void searches are safer since no digging is required.

C. Void searches are less likely to locate trapped victims than tunneling.

D. Victims in voids have a better chance of survival than do those buried in debris.

28. All of following are concerns of the officer in charge of a group performing selected debris removal except for which choice?

A. The officer should monitor the members' conditions, ensuring relief at intervals sufficient to avoid fatigue.

B. The officer should monitor progress and arrange for necessary support operations (such as changing worn-out saw blades) to ensure a smooth operation.

C. The officer should become closely involved with the manual labor, ensuring that the proper tools are being used correctly.

D. The officer should exercise extreme care and permit only specially trained personnel to move an object that may be supporting a large load.

29. When attempting to tunnel through debris to reach trapped victims, multiple avenues of approach are recommended, if possible. What are some considerations that you would use to determine whether multiple tunnels are practical?

A. staffing, equipment availability, and the stability of the debris

B. staffing, equipment availability, and the cause of the collapse

C. equipment availability, the stability of the debris, and the cause of the collapse

D. the cause of the collapse, the number of victims trapped, and the equipment available

30. When a variety of tools are available at a collapse that will all perform the same basic task, what considerations should determine which tools to use?

A. speed of operation
B. choice A plus exhaust fumes and noise produced
C. choice B plus sparks and vibration produced
D. choice C plus working room and monitoring equipment available

31. Which choice below does not accurately reflect a possible action to take if a wall or other object is threatening to cause a secondary collapse?

A. Shore the wall up.
B. Tie the wall back to a substantial object.
C. Pull it over into a safe area.
D. Remove the force acting on the wall.

32. Select the most correct statement regarding street management at the scene of a collapse.

A. Engine companies should position themselves at the nearest hydrants and prepare to supply master streams.
B. Tower ladders should be positioned in front of the building, as close as possible.
C. Ambulances should be staged in an EMS staging area with site access its main criterion.
D. Access routes must be maintained for heavy equipment, rescue units, etc.

33. All of the following precautions should be taken while operating at a structural collapse except for which choice?

A. Shut down all utilities, and monitor the atmosphere for flammable or toxic gases and sufficient oxygen.
B. Remove all personnel.
C. Control spread of fire or prepare for fire if none is present yet.
D. If you absolutely must cut a support, brace it, shore it, and prepare for secondary collapse; seek expert assistance.

23 FIRE DEPARTMENT ROLES IN TERRORISM AND HOMELAND SECURITY

Questions

1. Which of the following choices would not be a potential indicator of terrorist activity?

A. Tyvek protective suits and rubber gloves in a private house
B. laboratory-grade ice bath in a public storage locker
C. 500 lb of sodium cyanide in a chrome-plating factory
D. 500 lb of potassium cyanide in a farm outbuilding

2. What factors make American cities prime targets for terrorists?

1. large concentration of people
2. immediate worldwide media coverage
3. anonymity of the terrorists within the populace

A. all of the above
B. 1 and 2 only
C. 1 and 3 only
D. 2 and 3 only

3. An intelligence officer briefing fire department leaders on some of the criteria that should be considered in evaluating potential terrorist threats was reported to have made the following statements. The officer must have been misquoted in which statement, if his briefing was to be completely correct?

A. Easily recognizable landmark buildings are prime targets.

B. Financial institutions and corporate headquarters are also threatened.

C. Responses to crowded public locations should be approached with caution, especially if reports of chemical odor or persons overcome are received.

D. All locations where an explosion has occurred (except reported transformer explosions) should be treated as potential terrorist events.

4. The acronym CBRNE stands for what five categories of weapons?

A. chemical, bombs, reactive, nerve gas, explosive

B. chemical, biological, radiological, nuclear, explosive

C. chemical, bombs, radiological, nerve gas, explosive

D. chemical, bombs, reactive, nerve gas, explosive

5. Select the correct statement about chemical weapons:

A. Chemical weapons all involve military agents.

B. Chemical weapons are all fast acting and produce very visible displays of symptoms.

C. These agents can kill with extreme rapidity.

D. Fire department units are not needed at the scene of a chemical attack.

6. Select an incorrect statement concerning biological agents.

A. Biological agents are living organisms that cause diseases.

B. Initially, fire department units will most likely not know an attack has occurred.

C. Paramedic and ambulance personnel may be the first to detect an attack.

D. Advanced field detection equipment makes detecting biological agents relatively simpler than other agents.

7. A visiting lecturer made the following statements regarding nuclear and radiological threats to cities:

1. The detonation of a large nuclear device is a very serious threat and is considered very likely.
2. The detonation of even a very small nuclear device will totally wipe out most major cities.
3. The detonation of a radiological dispersal device or dirty bomb is the worst-case scenario.
4. A dirty bomb will fill the air with radioactive debris that can be deadly if inhaled.

A fire officer with limited knowledge of the threat of terrorism would know the lecturer was:

A. incorrect in all statements
B. correct in one statement
C. correct in two statements
D. correct in three statements
E. correct in all statements

8. Bombings are the most common type of terrorist attack using a CBRNE agent. Which statement regarding a bombing is incorrect?

A. Bombings pose the same risk to responders as other explosions.
B. At any explosion, the immediate priorities are lifesaving and control of the imminent hazards.
C. A rapid preliminary examination for likely cause is vital.
D. One of the most vital tasks at a bombing is dispersing crowds.

9. All of the following are correct actions to take at the scene of an explosion except for which choice?

A. rescue or evacuation of endangered persons and application of first aid to the injured; triage, treatment, decontamination if needed, and transport
B. monitoring of the scene to detect the possible presence of radioactive or chemical agents and potential perpetrators
C. extinguishment of fire and mitigation of other life-threatening hazards, gas leaks, hazmat releases, live power lines, etc.
D. primary and secondary searches of all affected structures; stabilization of collapsed/damaged structures to prevent further death or injury

10. Which of the following is not one of the potential hazards to be dealt with at an explosion scene?

A. fire
B. secondary explosions and ruptured gas, steam, water, or sewer lines
C. structural collapse including falling glass from surrounding buildings
D. hazmat release or exposure excluding biological hazards and bloodborne pathogens

11. In regard to modern terrorist bombings, which statement is least correct?

A. Most bombings are directed at property.
B. Extrication is not normally required at bus bombings.
C. These are large-scale mass casualty incidents.
D. Secondary attacks aimed at rescuers are increasing.

12. While you are responding to an explosion, which action would be least correct?

A. Watch for clues such as debris strewn on branches.
B. Note the actions of people in the area.
C. Avoid bottlenecks where an ambush is possible.
D. Position apparatus as normal for firefighting.

13. If you were to discover a bomb while searching for injured victims, which action would be least correct?

A. Immediately freeze in position, and do not touch anything.
B. Transmit a warning to all around you, and relay your findings to command via portable radio.
C. Evacuate at least 300 ft in all directions.
D. Notify law enforcement of the location and what you saw.

14. Company officers responding to an explosion would be correct if they did all but which choice?

A. relied on tank water for an attack
B. treated the entire area as a crime scene
C. treated the entire area as potentially contaminated
D. requested a hazmat unit to take samples of areas downwind of an explosion that has caused little damage

15. Among the items to be considered in sizing up an explosion scene, which item is least correct?

A. Determine the occupancy. Would it normally house materials that can cause an explosion?

B. Look at damage to the building, especially damage to wooden floors, columns, and girders.

C. Listen to what bystanders are saying, but always have some skepticism.

D. Listen to survivors, but always have some skepticism.

16. Regarding detonation of a radiological dispersal device (RDD), which statement is most accurate?

A. RDDs use ordinary explosives to detonate fissionable nuclear material.

B. In an RDD explosion, the main threats are trauma and radioactivity.

C. At large RDD explosions, radiation levels would approach those at a nuclear explosion.

D. Radiation levels would drop as you move farther from the site, especially moving upwind.

17. The role of the fire department at an RDD explosion is very important. What is one of the most important actions for first-arriving units to take?

A. Extinguish the fire in the debris, which sends more contaminated smoke skyward.

B. Use a solid stream to knock down radioactive plumes.

C. Keep all people out of the runoff water.

D. Use chemical detectors to survey the area.

18. Operations around radioactive materials rely on three key force protection factors to keep responders safe. What are they?

A. bunker gear, SCBA, and communications

B. bunker gear, SCBA, and shielding

C. bunker gear, SCBA, and radiation detection equipment

D. time, distance, and shielding

19. A firefighter wearing full bunker gear and SCBA would be most vulnerable to what type of radiation?

A. alpha particles
B. beta particles
C. cesium particles
D. gamma rays

20. Clues that a chemical attack is underway could include all of the following except which choice?

A. reports of a person unconscious
B. reports of chemical odors or fumes
C. presence of obvious explosive damage
D. biological indicators: people or animals showing signs or symptoms

PART III

FINAL EXAMINATION

24 FINAL EXAMINATION

Questions

1. Which elements of a coordinated fire attack will reduce or eliminate the life hazard?

1. removing all victims
2. venting to draw fire away from victims
3. confining the fire
4. extinguishing the fire

A. 1 only
B. 1, 2, and 3 only
C. 1, 3, and 4 only
D. all of the above

2. What items from the size-up combine to produce the life hazard?

1. location and extent of fire
2. time of day
3. occupancy
4. height
5. weather
6. apparatus and personnel

A. all of the above
B. 1, 2, and 3 only
C. 2, 3, 5, and 6 only
D. 1, 3, 4, and 5 only

3. A captain cited the following items as being major concerns about a building's construction that would affect the size-up. Which were correctly cited?

1. degree of compartmentation
2. combustibility of the building
3. number of hidden voids
4. ability of material to resist collapse

A. 4 only
B. 2 and 4 only
C. 2, 3, and 4 only
D. all of the above

4. According to the NFPA (National Fire Protection Association), which occupancy has been shown to be the most deadly per incident to firefighters in the period 2004–2008?

A. factory fires
B. residential fires
C. public assembly fires
D. store fires

5. The preferred method of attack on a fire in the incipient stage is listed in which choice?

A. direct attack
B. indirect attack
C. combination attack
D. z-attack

6. While waiting for water, the nozzle team should look along the floor of the fire area. They are looking for:

1. possible victims
2. location of the fire (glow)
3. layout of the area
4. location of utilities

A. 1 only
B. 1 and 2 only
C. 1, 2, and 3 only
D. all of the above

7. The indirect method of attack may be suitable for use on third stage fires. All of the following are required ingredients to make this attack successful except which choice?

A. high heat condition present
B. limited ventilation
C. limited number of occupants
D. limited size of potential fire area

8. What strategic option is *required* at a two-bedroom fire in an occupied apartment house?

A. indirect attack
B. offensive attack
C. direct attack
D. defensive attack

9. At 3 o'clock in the morning, you arrive simultaneously with the first engine to find fire venting from two windows on the ground floor of a three-story brick and wood joist warehouse. The fire is exposing a nearby fireproof cold storage warehouse. Where do you order the first line to be placed?

A. between the fire building and the exposure
B. inside the exposure
C. inside the fire building
D. in front of the fire building, supplying a master stream

10. **List the following selection of victims in the priority order of their removal, assuming a serious fire condition on the second floor of a six-story, 100×150-ft, brick and wood joist apartment house.**

    1. persons on the top floor
    2. persons on the fire floor, remote from the fire
    3. persons on the floor directly above the fire
    4. persons on the fire floor, in proximity to the fire
    5. persons on the floor below the fire
    6. persons on the fourth floor

    The correct order is:

    A. 4, 3, 2, 6, 1, 5
    B. 4, 2, 3, 6, 1, 5
    C. 4, 3, 1, 6, 2, 5
    D. 4, 3, 1, 2, 5, 6

11. **All of the following situations pose special firefighting problems due to their location except for which choice?**

    A. cellar fire in a fireproof office building
    B. top-floor fire in a frame apartment complex
    C. fire on the top floor of a three-story, windowless warehouse
    D. second-floor fire in a four-story fireproof hotel

12. As the incident commander, you arrive simultaneously with the first engine at a working fire in an old downtown three-story brick and wood joist building. The fire seems to be located in the first-floor office supply store, with no obvious extension to the occupied apartments above. As the engine stretches its line to the front door, you note that the show windows are pitch black, the glass is cracked, and heat radiates from them as you pass. There is no open flame visible, but heavy smoke puffs from around the door, then seems to be drawn back inside. At this point your most appropriate order is which choice?

A. Order the engine company to take a position across the street from the fire building for possible use of its master stream.

B. Order all the show windows vented immediately after the engine gets water in its line.

C. Order the engine to use an indirect attack with wide fog after all the show windows are vented.

D. Order the ladder company members to withhold venting of the upper floors until the ground floor fire is extinguished.

13. What factors determine whether a hose stream will be capable of extinguishing a given structure fire?

1. type of nozzle: fog or solid
2 reach of stream
3. volume of stream
4. mode of operation

A. 1 and 2

B. 2 and 3

C. 3 and 4

D. 1, 2, and 4

14. Where should the first line be placed at a serious cellar fire in an occupied house that has vented out the cellar window and is now blow-torching up the asphalt siding on the rear wall?

A. at the outside cellar doorway on the side of the house

B. at the rear cellar window to knock down the wall and operate into the cellar

C. between the fire building and the most serious exposure

D. through the front door to the interior cellar stair

15. All of the following are prerequisites for high-expansion foam to successfully knock down a cellar fire except for which choice?

    A. The fire must involve only Class A materials.
    B. The floor above the fire must be vented.
    C. Sufficient volume of foam per minute must be applied.
    D. The foam must be able to reach the seat of the fire.

16. How much hose should be stretched for a fire in the far rear corner of the top floor of a four-story, 100×100-ft apartment with the hydrant 200 ft from the front entrance? The stair is a return stair, just inside the front door and has no well hole.

    A. 550 ft of 1½-in.
    B. 550 ft of 1¾-in.
    C. 550 ft of 2½-in.
    D. 300 ft of 1¾-in. attached to 250 ft of 2½-in.

17. A senior firefighter in an engine company made the following statements concerning hoseline positioning to a group of newer members. The captain, overhearing the firefighter's comments, would have been wise to correct the member in which statement?

    A. For fires above grade, stretch dry hose to a safe area, usually via the interior stair.
    B. For fires above grade, it may be beneficial to haul the line up onto the fire floor by using a rope.
    C. For fires below grade, always charge and bleed the line before starting down the stairs.
    D. For fires in one-story buildings, place the hoseline between the fire and the occupants.

18. What effect would doubling the length of a hoseline have on the total friction loss if the flow remained the same?

    A. no effect
    B. one-half the loss
    C. double the loss
    D. quadruple the loss

19. Where long distances and large flows are required of a relay operation, which choice does not solve the water-delivery problem?

    A. laying multiple supply lines
    B. using an automatic nozzle to use the available water most efficiently

C. using larger diameter supply lines
D. using a larger capacity pumper at the source

20. An engine company officer arriving to find heavy fire in control of the front 25% of a 100×100-ft lumber storage building should know that in order to knock this volume of fire down, her crew will have to attack with most nearly _______ gpm.

A. 250
B. 1,250
C. 2,500
D. 5,000

21. The captain of a ladder company, while conducting training in the quarters he shares with an engine company, made the following statements regarding engine company operations. In which choice did the esteemed captain misstate a fact?

A. When positive pumping through large-diameter hose, the pumper stops at the fire and drops off a variety of hose and nozzles, a manifold, and appliances, then proceeds to a water source.
B. By placing the manifold near the fire, a variety of options are available.
C. A manifold might be used to supply handlines, master streams, or other pumpers.
D. The manifold should supply only one class of device, but if that class is handlines, there is no problem.

22. Serious fires in strip malls can best be dealt with using elevating platforms. All of the following illustrate good use of an elevating platform except which choice?

A. members in the basket removing a sign on the front of an old taxpayer to expose the cockloft
B. using a power saw from the basket to open the fascia of the overhang in front of the stores
C. using the platform nozzle from the sidewalk level to direct its stream into the cockloft
D. using the platform nozzle from above the roof level to extinguish fire coming through holes in the roof

23. What is the most common cause of false alarms involving automatic dry-pipe sprinkler systems?

A. pressure surges in supply
B. loss of water pressure

C. loss of air pressure

D. malicious mischief

24. All but one of the following are locations where automatic dry-pipe sprinkler systems would be appropriate protection. Which choice does not belong?

A. over the offices in an ice cream factory

B. over the freezer area in an ice cream factory

C. over the loading dock at a northern furniture factory

D. over the cars at a multilevel parking garage in Canada

25. A unit arriving at a large warehouse at 3 a.m. on a cold February morning (0°F) in response to a sprinkler valve alarm noted the following: no alarm bell sounding, no water discharging from the drain piping, and no visible smoke or fire. The alarm company advises them that the valve has not reset. The most correct action would include which choices?

1. Return to quarters immediately.
2. Remain on scene awaiting the plant foreperson.
3. Force entry and began search of premises.
4. Send a knowledgeable member to locate the controls.
5. Have the member at the controls immediately shut down the system that had activated.

A. 1 only

B. 2 only

C. 3 only

D. 3, 4, and 5

26. A visiting chief lecturing about high-rise firefighting techniques made the following statements. He should be corrected in which choice?

1. A separate pumper should be used to supply each Siamese.
2. If the Siamese is damaged, or if sufficient pressure cannot be developed, hoselines should be stretched to pump into lower floor standpipe outlets.
3. The lower floor hose outlets should be supplied at any serious fire as a precaution against total water loss should the Siamese or its piping fail.
4. Standard fog nozzles operated from standpipe systems should be supplied at an engine pressure of 100 psi.

A. none of the above
B. 2, 3, and 4 only
C. 3 and 4 only
D. 4 only

27. Under what circumstances should fire departments shut down sprinkler valves during a serious fire?

1. where explosions have demolished sprinkler piping
2. where fire department handlines are in place to assume operations
3. where the fire has already overwhelmed a sprinkler system and the water supply is needed to protect exposures
4. where firefighters are unable to locate the fire due to the smoke and water spray

A. all of the above
B. 1, 2, and 3 only
C. 1 and 3 only
D. 1 only

28. During a fire in a high-rise building, who should be responsible for maintaining the proper pressure from the standpipe outlet?

A. the first-arriving engine chauffeur
B. the officer in command of the hoseline
C. the officer in command of the fire
D. a radio-equipped member stationed at the outlet

29. Firefighters respond to a sprinkler alarm at a warehouse at 2 a.m. and are advised that the alarm has reset itself. Which type of sprinkler system is most likely present in this building?

A. automatic dry pipe
B. automatic wet pipe
C. nonautomatic
D. deluge

30. Which of the following are advantages of using a front porch roof as a platform for VES?

1. It provides a stable work platform.
2. It provides a safe area of refuge if the member is forced to retreat.
3. Assistance is generally available in the front of building.
4. Being on the outside of building, the member does not need an SCBA.

A. all of the above
B. 1, 2, and 3 only
C. 1 and 2 only
D. 1 only

31. Choose the incorrect statement below concerning the proper positioning of aerial ladders and platforms for rescue where time is crucial.

A. For a platform, stop the apparatus with the basket in line with the objective.
B. Position the basket with the top rail just below the windowsill, to allow easy access through the basket gate.
C. Aerial ladders should be positioned with the turntable in line with the objective.
D. The tip of the aerial ladder should be positioned just below the windowsill.

32. A chief officer lecturing several of her members about overhaul made the following statements. She should have been tactfully corrected for which statement(s)?

1. Vertical voids, particularly those housing water, sewer, and vent pipes, as well as those for steel I-beams, must be the first priority when checking for extension.
2. Once vertical extension is checked, horizontal voids must be examined.
3. Horizontal voids are only found on the top floor.
4. At an old wooden home, be sure to quickly open all the baseboards on the interior walls of the floor above the fire.

A. all of the above

B. 2, 3, and 4

C. 3 and 4 only

D. 4 only

33. A chief officer is faced with a serious fire on the top floor of a four-story, 75×100-ft apartment house that appears to have extended into the cockloft. There are no adjacent exposures. The department's sole aerial device, a 75-ft snorkel, is heavily engaged in rescuing trapped occupants. The chief transmits the following calls for assistance. Select the least correct choice of devices.

A. one additional snorkel to rescue other trapped victims in the rear

B. yet another snorkel to the rear for an escape route for the many firefighters involved in roof operations

C. an aerial ladder to the front to provide roof access

D. a ladder tower for possible use of its heavy stream

34. A lieutenant of a ladder company was lecturing his members and the members of a neighboring ladder company on positioning their apparatus for most effective use. He was most correct in which statement?

A. Slow down before reaching the building and observe conditions. If a rescue is needed, spot at the best location.

B. If the unit is a midship-mounted device, spot the apparatus at about a 45° angle away from the building.

C. For rear-mounted devices, the best scrub area is obtained by nosing the apparatus in toward the building.

D. When approaching, if no obvious need is seen for the device, stop the rig about 15 ft short of the far end of the building.

35. **A lieutenant lecturing her members on the use of through-the-lock forcible entry makes the following statement: "Use of the through-the-lock method is indicated by ________________."**

    1. heavy fire condition
    2. specific types of doors
    3. specific types of locks

    Which choice(s) correctly completes the statement?

    A. all statements
    B. 1 and 3 only
    C. 1 and 2 only
    D. 2 and 3 only

36. **A hydraulic forcible-entry tool should be used in place of conventional or through-the-lock methods in all but which choice?**

    A. where speed is essential
    B. in poor visibility
    C. on outward-opening doors
    D. where damage is not a major concern

37. **All but one of the following are useful tools in forcing case-hardened padlocks. Which choice does not belong?**

    A. duckbill lock breaker
    B. power saw with aluminum oxide blade
    C. Halligan tool
    D. 30-in. bolt cutters

38. **The American Lock 2000 series gate lock may be removed using which tool?**

    A. duckbill lock breaker
    B. power saw with aluminum oxide blade
    C. Halligan tool
    D. 30-in. bolt cutters

39. Advantages of horizontal ventilation over vertical ventilation at most structure fires include all but which choice?

A. It is faster and easier than roof cutting.

B. It requires less personnel.

C. It provides superior smoke movement.

D. It is less costly to repair.

40. What is the most serious danger to a firefighter cutting a vent hole on a metal deck and bar joist roof?

A. falling into the hole he is cutting

B. being caught when large sections of the roof suddenly fail

C. total roof failure in less than five minutes of fire exposure

D. sudden failure of highly heated joists struck by hose streams

41. A trench cut would be an effective defensive measure in which situation?

A. H-type apartment building with a serious cellar fire

B. serious cockloft fire in a large supermarket

C. heavy fire blowing out of the middle store in a row of 12 taxpayers

D. long, narrow, one-story warehouse with fire at one end

42. Which choice lists a factor that is not an advantage of positive pressure ventilation (PPV)?

A. PPV provides more efficient air movement.

B. The PPV fan does not clutter access to building.

C. PPV can safely remove flammable gases and vapors.

D. PPV works best on advanced fires that have self-vented from many openings.

43. When encountering a heavy fire in a building with energy-efficient windows that have not yet vented, when should horizontal ventilation not be performed?

A. immediately, while the attack team is in a safe area behind a closed door

B. after the attack hoseline has thoroughly cooled the area

C. just before the nozzle team opens the nozzle

D. upon reaching the windows from the inside, behind the nozzle

44. As the officer in command of the roof sector, you are informed that the roof you are working on appears to be a rain roof. Given this bit of information, you would be most correct in which statement?

A. The biggest problem will be determining the extent of the fire in the cockloft.
B. Roof stability is not a major concern if, after you make an inspection opening, there is no fire directly under your position.
C. Roof ventilation will be very rapid, given the nature of the roof.
D. Once the rain roof is opened, push down the ceilings below to expose the cockloft.

45. Which choice is not true regarding lightweight parallel chord trusses?

A. They create large open spaces for fire to travel in.
B. They can be recognized by their classic hump shape.
C. Operating a power saw on the deck they support risks causing total truss failure.
D. Failure of a single web member or any of the connectors (gusset plates) can cause total truss failure.

46. While en route to an alarm for a fire in a row of garden apartments, your dispatcher informs you that the preplan describes the building as being built with lightweight parallel chord trusses for floor as well as roof support. Given this information, and a reliable report that all occupants are out of the fire apartment, what would not be an acceptable tactic?

A. Operate from the exterior, using the reach of the stream on the contents.
B. Provide total ventilation directly over the fire.
C. Open the side wall from a platform and direct a stream into the truss loft.
D. Advance slowly after the fire is knocked down, and use adequate lights, ventilation, and shoring.

47. A ladder company officer is directed to supervise the operations on the roof of a well-involved commercial building. Upon reaching the roof, she attempts to determine what type of roof the unit is working on. She would be correct in ordering an immediate evacuation of the roof if she discovered which type of roof?

A. a 2×4 inverted roof, with ¾-in. plywood deck
B. a standard sawn joist roof, with 1×6 roof boards
C. a concrete deck roof on protected steel I-beams
D. a wooden I-beam supported plywood roof deck

48. All but one of the following are correct actions to take when cutting a trench on an H-shaped apartment house. Which choice is incorrect?

A. The trench is cut close enough to limit fire extension, yet far enough away to allow completion without being passed.

B. The trench should be cut where the building profile is at its widest.

C. The trench is located in the path of anticipated fire travel.

D. The trench must be reinforced with hoselines.

49. For a serious top-floor fire in a multistory, flat roof structure, what is the proper sequence of actions to be taken by the roof crew?

1. Vent over the stairs.
2. Cut the roof over fire.
3. Vent top floor windows.
4. Vent over other shafts.

A. 1, 2, 3, 4

B. 1, 3, 2, 4

C. 1, 4, 3, 2

D. 2, 1, 4, 3

50. A ladder company officer, discussing roof-cutting operations with his members, made the following statements. He should be corrected in which choice?

A. The wind direction is the first item to consider when deciding where to cut.

B. Plan an escape route in case conditions deteriorate.

C. Never cut so that a member has to step on a compromised portion of the roof.

D. Cut close to the inside of the joists when making the traditional 8×8-ft hole.

51. Which choice below would not be part of the secondary search of a structure?

A. the outside vent person searching in bushes and shrubs outside a house

B. the irons person checking inside a kitchen cabinet during overhaul

C. the nozzle operator looking in along the floor while waiting for water

D. the can person checking inside each drawer in a dresser while performing salvage work

52. All but one of the following descriptions of a team search are correct. Which choice is incorrect regarding a targeted fan search?

A. Each team search needs a leader, an anchor, either two or four searchers, a control person at the entrance, and a two-person rescue team.

B. The control person pays out the desired amount of rope while the team advances.

C. The searchers are attached to the main line by 25-ft lengths of rope.

D. The searchers fan out from the main line and move forward until they meet at the anchor's location.

53. Where should you begin searching when entering the area directly above the fire?

A. moving quickly toward the fire, then working back toward the door

B. inside the room directly over the fire

C. around the perimeter of the entire building

D. at the door to the area, then working in toward the fire

54. As the incident commander, you find yourself confronting a heavy fire on the ground floor of a four-story, brick and wood joist row house, with one apartment per floor. What is the desirable minimum number of personnel you should assign for search purposes?

A. 2

B. 6

C. 8

D. 10

55. A firefighter who finds herself lost in a large, smoke-filled warehouse, where the fire is being held in check by sprinklers, should take all of the following actions except for which choice?

A. Move rapidly while searching for an exit.

B. Call for help, by voice, radio, and PASS.

C. Use all means at her disposal to attract attention to her position.

D. Lie with her face close to the floor and turn off her light for a few moments to look for other light.

56. A company officer discussing the firefighters survival survey with the company members made the following comments concerning "What is happening to the building?"

1. You must constantly be evaluating the potential for flashover.
2. You must be evaluating the potential for a backdraft.
3. You must be aware of fire traveling in voids around you.
4. You must be aware of the ceiling height above you.
5. You must monitor the stability of the building.

This officer was correct in which choice(s)?

A. 5 only
B. 3, 4, and 5
C. 1, 2, 3, and 5
D. all of the above

57. At a minimum, when should a roll call be ordered?

1. when a PASS alarm has been activated for more than 30 seconds
2. after the collapse of a member from an apparent heart attack
3. after sudden fire extension
4. after an emergency evacuation from a building

A. all of the above
B. 1, 2, and 3 only
C. 1, 3, and 4 only
D. 1 and 4 only

58. What are "Norman's rules of survival on the fireground"?

1. Always have a charged hoseline before entering a burning building.
2. Always know where your escape route is.
3. Always know where your second escape route is.
4. Never go above the fire.
5. Never put yourself into a position where you are depending on someone else to come get you out.

A. 1, 2, and 3
B. 2, 3, and 5
C. 2, 4, and 5
D. all of the above

59. A firefighter who finds herself entangled in cable TV wires while searching should not perform which action?

A. Inform her partner of the difficulty.
B. Remove the facepiece to untangle it from the wires.
C. Try backing up, dropping lower, and proceeding forward.
D. Perform the emergency escape maneuver.

60. In which type of home would a cellar fire mandate a hoseline to the top floor or attic?

A. platform frame home
B. braced frame home
C. bubble frame home
D. balloon frame home

61. Which style of roof cutting is preferred for attic fires in plywood-covered peaked roofs?

A. the strip cut at the peak
B. the quick cut
C. the basket cut
D. the standard 4×4-ft cut

62. Once fire is found spreading up a balloon frame wall, speed is of the essence. Which of the following is a good action to take in this event?

A. If high heat or fire is present, drive the hose stream up the bay.

B. Position a hoseline on the top floor and one in the cellar.

C. Quickly expose each bay for its entire length, especially bays above and below windows.

D. Do not bother with roof ventilation if the fire is in the cellar.

63. What is an unacceptable use of an elevating platform at a serious fire in an occupied apartment house?

A. hauling a hoseline up to the fifth floor (the fire floor)

B. delivering power saws and people to the roof

C. supplying a handline from a basket outlet on the floor above the fire

D. using the master stream to knock down fire in the cockloft

64. Your unit is ordered to search the floor above the fire in a large apartment house. As you ascend the interior stair past the fire floor, you notice the fire is definitely not yet under control. You also receive a radio report of persons trapped and jumping from the floor above. In this case, it would be best for you to do which choice?

A. Delay ascending the stair until the fire is under control.

B. Ascend immediately and force entry directly into the apartment right over the fire apartment.

C. Delay ascending until a charged hoseline is brought to the floor above.

D. Ascend immediately, but force entry into one or more apartments other than directly over the fire, for use as areas of refuge if the hall becomes impassable.

65. While conducting the primary search of the floor above the fire in an SRO, you come across a number of doors that are padlocked on the hall side. You should do which of the following choices?

A. Force them all and search in each for occupants.

B. Skip past them as there is no life hazard here.

C. Force them, but move on; a later unit will search them.

D. Search these areas only after obtaining the keys for each.

66. A lieutenant is discussing fires in Class 1 high-rise apartment buildings with his crew when he makes the following statements. He was correct in all but which choice?

A. When confronted with high wind conditions, you must evaluate the effect that venting windows will have.

B. Under these conditions, if you are venting from inside the fire apartment, make a small experimental opening and see what effect it has on the fire.

C. If venting from the floor above, be sure the door to the stairwell is closed.

D. If the wind blows in when the windows are opened, it is best to withhold venting until the hoseline has thoroughly cooled the area.

67. A chief officer arrives at a fire in a high-rise residential building to find the wind blowing the fire back into the building. Upon opening the door from the stairway into the public hall, the firefighters are blasted back down the stairs by a blowtorch of flame. Under these circumstances, what should be the last thing the chief orders his companies to try?

A. Push into the hall with two handlines simultaneously from the same stair.

B. Breach a hole from the stairway into the adjacent apartment, and advance within it to a point where you can put water on the fire.

C. Use positive pressure fans to push the fire back into the apartment and out the windows.

D. Use an outside stream to darken down the fire.

68. At a serious fire on the ground floor of a four-story ordinary construction apartment house, two members immediately ascended to the roof via aerial device to vent over the staircase. After completing this duty and other ventilation of vertical arteries, these members should perform which task?

A. Descend the interior stairs and search the top floor.

B. Feel soil pipes that line up vertically with the fire area for heat.

C. Reach over the parapet and vent all top floor windows with a hook or pike pole.

D. Immediately begin cutting the roof.

69. Fires involving garden apartments or townhouses that are in the open framed stage of construction are best attacked with which hose stream?

A. a 2½-in. handline equipped with a 1¼-in. solid tip

B. a single 1¾-in. line through the front door

C. two 1¾-in. lines operating side by side

D. a preconnected master stream

70. Which statement correctly describes the features of a garden apartment or townhouse?

A. Garden apartments are single-family dwellings.

B. Townhouses are multiple dwellings.

C. Garden apartments may have common basements or crawl spaces.

D. Nearly all townhouses have common cocklofts.

71. Which statement does not correctly describe proper tactics for use at a fire in townhouses and garden apartments?

A. Protection of the interior stair is critical at a townhouse fire.

B. Protection of the interior stair is critical at most garden apartment fires.

C. The preferred hoseline for apartment fires is the 1¾-in. line.

D. The preferred hose stretch for fires out of the reach of the preconnect is a lead of 100 ft of 1¾-in. line attached to a break-apart nozzle on a 2½-in. line.

72. Which of the following structural elements would not commonly be found in a new strip mall?

A. plasterboard walls

B. concrete slab floors

C. corrugated metal decks

D. wooden roof joists

73. Commercial buildings have all of the following items in common with nonfireproof multiple dwellings except for which choice?

A. common cockloft over entire building

B. large floor areas

C. built with ordinary construction

D. high life hazard regardless of the time of day

74. **Under which set of circumstances would it be proper to smash the front plate glass windows of a row of commercial occupancies?**

    A. a light haze present in the middle store of a row of 13 stores
    B. a store that is puffing heavy smoke with no visible fire
    C. in the third store in a row that has fire visible, extending via the cockloft from two other fully involved stores
    D. at a store at 3 a.m. where the windows are already blackened and cracked from the intense heat inside

75. **A serious cellar fire in all but which occupancy should cause the OIC to consider early collapse potential?**

    A. an old-fashioned soda fountain with terrazzo floor
    B. a pizza parlor with an asphalt tile floor
    C. a self-service laundry
    D. a drugstore with a ceramic tile (mud) floor

76. **The danger of ceiling collapse in taxpayers is most prevalent in which structures?**

    A. those newly built with lightweight truss roofs
    B. newer Class 2 (noncombustible) taxpayers
    C. newer Class 3 (ordinary) construction taxpayers
    D. older taxpayers with multiple hanging ceilings

77. **Engine companies arriving at working fire at a strip mall performed the following actions. The incident commander needs to speak to which company officer regarding the officer's choice of tactics?**

    A. Engine 1 laid 200 ft of 5-in. line and positioned the apparatus directly across the street from the fire store.
    B. Engine 2 connected to a hydrant with its 6-in. soft suction and fed three handlines.
    C. Engine 3 laid in 400 ft of 3-in. line to the rear alley to protect exposures.
    D. Engine 4 laid two lines of 3-in. hose from Tower Ladder 2 and connected to a hydrant around the corner.

78. Which of the following are warning signs of potential backdraft?

1. heavy smoke, issuing under pressure
2. highly heated windows
3. smoke puffing out, then being drawn back in
4. large amounts of visible flame

A. all of the above
B. 1, 2, and 3 only
C. 1, 3, and 4 only
D. 1 and 3 only

79. Despite all efforts to advance a handline into a cellar fire of an old downtown business, no progress is being made after 10 minutes due to high heat conditions. What orders would you as the incident commander give next?

A. Deploy fresh personnel to renew efforts.
B. Withdraw handlines from the cellar and place distributors or cellar pipes through holes cut in the wooden first floor.
C. Introduce high-expansion foam through the cellar stairs.
D. Flood the first floor with master streams.

80. A chief officer, arriving at a working fire in a newer strip mall, finds fire venting from the end store in a row of 14 stores. The roof team reports the roof is sagging over the fire store. What would be the most correct order for the chief to give?

A. Advance a handline into, and vent the roof over, the involved store.
B. Advance a handline into an exposed store 60 or more feet ahead of the fire, and begin sweeping the cockloft with the stream. Have the roof team drop back to this store.
C. Evacuate the roof as well as the interior of the building, and conduct only exterior operations.
D. Evacuate the roof only, and advance the handline into the fire store, sweeping the ceiling.

81. What creates the greatest threat to civilian life in post–World War II high-rise office buildings during fires?

A. lightweight construction techniques that lead to collapse

B. large, open floor spaces

C. inability to rapidly perform effective ventilation

D. use of core construction techniques

82. During a serious high-rise fire, a crew of firefighters was sent to the roof level. There they encountered the roof bulkheads of three separate staircases. They would be most correct if they withheld venting which stair immediately?

A. the evacuation stair

B. the ventilation stair

C. the attack stair

D. none of the above

83. At a serious fire on the 33rd floor of a high-rise office building, a company officer made a number of statements regarding elevator use. Which is the most correct statement?

A. Firefighters' service elevators should always be used, since they are specially designed to protect firefighters in the event of an emergency.

B. If firefighters' service elevators are not present, use freight or service elevators for their larger load capacity.

C. Each team that enters an elevator must include masks for each member, a radio, and a set of forcible-entry tools.

D. Upon entering an elevator car, immediately press the "call cancel button," then take the car directly to the floor below the fire.

84. Two fire officers working in a downtown metropolitan fire station made the following statements about stack effect. They should have been corrected in which choice?

A. Stack effect is most pronounced in tall buildings.

B. Stack effect during a top-floor fire on a cold day can be an extreme problem.

C. Stack effect during a lower-floor fire on a cold day can be an extreme problem.

D. Stack effect during a top-floor fire on a hot day can be an extreme problem.

85. Which of the following types of buildings poses the most severe danger of collapse in the event of a serious fire during the construction of the building?

A. ordinary brick and wood joist building

B. precast concrete building

C. steel frame building with spray-on fireproofing

D. poured-in-place, reinforced concrete building

86. An engine company arrived first at an apartment house on a bitterly cold night, in response to a report of an explosion in the cellar, and found the entire cellar knee-deep in a ghostly white mist that smells strongly of fuel oil. They immediately took the following actions, all of which were correct except for which choice?

A. evacuated the entire building

B. immediately entered the cellar to shut down the oil at the tank valve

C. stretched a handline with a fog nozzle and saturated the cellar

D. removed all sources of ignition and vented the area with a positive pressure fan

87. Early on a still summer evening your company arrives first-due at a crowded new residential development. You find a fire in a barbecue involving a 20-lb propane tank in the rear of one of the homes. You issue the following orders, but which one should you correct?

A. Secure a water supply, and stretch a 1¾-in. line to cool the cylinder.

B. You have the second engine stretch another line to the interior of the home to check extension there.

C. You order the homes on each side and to the rear evacuated.

D. Once the lines are in place, you order the nozzle operator to extinguish the fire and then shut off the cylinder valve.

88. When using the combination attack on a free-burning fire involving several rooms of a home, the nozzle operator performed the following actions. Which was not a proper action?

A. used a very narrow fog pattern, directed at the ceiling

B. moved the stream in a side-to-side sweeping motion along the ceiling

C. continued this sweeping motion along the ceiling as the line was advanced from room to room

D. after the fire had darkened down, shut down the nozzle to give the smoke, heat, and steam a chance to lift

89. A firefighter finds herself cut off by a rapidly extending fire. This member would not be operating in her own best interests if she performed the actions found in which choice?

A. stayed calm and stayed put, did not move around so as to conserve air

B. called for help by all means possible

C. closed doors between her and the fire

D. activated her PASS device

90. A chief officer in command of roof operations at a large fire in the cockloft of an H-type building is supervising a trench cut operation. He gave the following orders. This chief should probably be demoted if this is an indication of his tactical prowess. He was correct in which choice?

A. Make the trench 2 ft wide from fire wall to fire wall.

B. Make sufficient inspection holes on both the fire side and the safe side of the trench.

C. The trench must be subdivided into 2-ft segments.

D. Don't begin pulling the trench until fire shows at the inspection holes on the safe side of the trench.

91. A ladder company officer arrives at the scene of an early morning fire in a three-decker wood-frame apartment building that has one apartment on each of its three floors. Fire is venting out of two windows on the front of the second floor. Due to another nearby multiple-alarm fire, there will be a long delay in the arrival of the second ladder company. Which of the officer's actions are inappropriate under these circumstances?

A. transmitted calls for additional assistance

B. while searching the fire apartment, began the search at the apartment door, then worked her way toward the seat of the fire

C. after completing the primary search of the fire floor, ascended to the floor above to search it

D. began searching the top floor at the apartment door and moved toward the front of the apartment, where there was a fire escape

92. An engine company officer arrives to find one three-story frame building fully involved with fire and a similar building located 30 ft upwind smoking from the radiant heat. This officer would be acting most correctly in which choice?

A. directed the driver to position the apparatus to use the apparatus deck pipe on the fire building as well as the exposure

B. had the deck pipe used as a water curtain between the fire building and the exposure

C. had a handline positioned to coat the fire building with water to absorb the radiant heat

D. used another handline with a solid tip nozzle to assist the deck pipe in cooling the exposure windows, since the handline could get very close to the building

93. Your unit is part of the assignment dispatched for an odor of smoke in a department store, just after closing time. On arrival, you are met by the building manager, who tells you there is a strong odor of smoke on all three floors of the building. Investigating, you smell the odor, but detect no visible smoke. What is the most correct action for you to take as the ranking officer at the scene?

A. Return all units, and advise the owners to call back if they see smoke.

B. Survey incandescent light fixtures for ballasts that might be overheated.

C. Move to where the odor was last detected and examine appliances and devices in that vicinity.

D. Send a team to the machinery rooms and examine the HVAC and elevator motors for overheating.

94. At a recent fire in an older supermarket, the high heat and heavy smoke from the cockloft in the rear of the store prevented handlines from advancing to the seat of the fire. The chief ordered the lines withdrawn to the sidewalk in anticipation of full involvement. The chief's actions were:

A. Appropriate. The inability to advance makes full involvement inevitable.

B. Inappropriate. The lines should have been relieved with fresh personnel and advanced to the seat of the fire.

C. Appropriate. The handlines will be operating in better conditions outside.

D. Inappropriate. The sidewalk is within the collapse zone.

95. A firefighter operating at an advanced fire in a wood-frame home watches as a door suddenly swings from the open to the closed position, for no obvious reason. This warning sign should most likely indicate to the member that he should evacuate the building because of the possibility of:

A. backdraft

B. flashover

C. structural collapse

D. poltergeist explosion

96. A captain lecturing firefighters about collapse rescue on the fireground made the following statements. She was correct in which choices?

1. In the event a member is trapped, all members should immediately come to assist in the member's removal.
2. All firefighting and support functions should cease, and those resources used to rescue the trapped members.
3. All equipment is considered expendable when attempting a firefighter's rescue.
4. Once all victims are accounted for, shift back to a very cautious approach; be sure everyone is informed.

A. all of the above

B. 2, 3, and 4 only

C. 3 and 4 only

D. 4 only

97. A chief officer arriving at the scene of a large building that has collapsed pancake fashion ordered the following actions. In which choice should the chief reconsider his actions?

A. He ordered the first- and second-arriving engines to take hydrants in the block and supply handlines or master streams.

B. He ordered aerial ladders to take positions away from the front of the collapse site.

C. He ordered an elevating platform called, to be positioned in front of the collapse site, outside the secondary collapse zone.

D. He ordered space reserved near the collapse scene for several heavy rescue units, which he called for special equipment.

98. All of the following safety precautions should be taken at the scene of a structural collapse, except for which choice?

A. Control the spread of fire; if no fire is present, prepare for it anyway.

B. Eliminate vibrations from nearby highways, rail lines, nonessential apparatus, etc.

C. Use all available personnel for maximum impact.

D. Monitor the atmosphere for oxygen, flammable, and toxic gases.

99. A firefighter wearing full bunker gear and SCBA would be most vulnerable to which type of radiation?

A. alpha particles

B. beta particles

C. cesium particles

D. gamma rays

100. In the current world environment, where there exists a potential use of weapons of mass destruction against civilians and first responders, incident commanders must be able to recognize situations where committing fire department personnel may be placing them in grave danger. This may require the department to take strictly defensive positions and actions. Which choice below does not illustrate a situation where only defensive actions should be taken?

A. First-arriving units are already casualties, felled by an unknown agent in spite of wearing full personal protective equipment.

B. Secondary attacks have occurred or are ongoing.

C. No more viable victims can be seen or heard.

D. More than 10 minutes has passed since dispersal of a chemical agent.

101. You arrive as the officer in command of the first-due engine at a fire in a four-story apartment house at 3 a.m. You find a fire venting from three ground-floor windows and heavy smoke showing from all floors. Your crew consists of a driver, one firefighter, and yourself. The next unit to arrive will likely be 5–7 minutes behind you. Numerous bystanders are reporting multiple people trapped in the 16 apartments in the building. All of the following are incorrect actions to order, except for which choice?

A. Order the unit's portable ladder raised to remove the trapped occupants.

B. Order the hoseline placed to protect the interior stairs and confine the fire.

C. Order the members to begin search efforts on all floors.

D. Order the members to begin ventilating the structure.

102. **Which of the following conditions is not likely to require a special hosebed on an apparatus to allow rapid stretching of a handline of unusual length?**

A. a sewer trench
B. street construction
C. deep snow
D. a large high-rise building

103. **What size-up factors combine to produce the life hazard at a structural fire?**

1. time of day
2. occupancy
3. location of fire
4. extent of the evacuation

A. all of the above
B. 1, 2, and 3 only
C. 1 and 2 only
D. 2 only

104. **Several unusual circumstances pose serious dangers to firefighters attempting to gauge the collapse potential at Class 5 buildings. All of following are correctly listed except for which choice?**

A. Fire was set with accelerants.
B. Building has experienced previous fires.
C. Building has undergone extensive renovations.
D. Building was constructed with fire-resistive construction.

105. **An engine company arriving at a strip mall fire to find one 20 x 75 ft. store fully involved would be acting correctly by using all of the following streams except for which choice?**

A. a 2½-in. handline
B. a preconnected deck gun
C. two 2½-in. handlines
D. a portable master stream blitz line

106. Which is an incorrect statement made about the nozzle team's activities while it is waiting for water?

A. The firefighters should stay low and out of the doorway.

B. The nozzle operator should bleed the line of air and check the pattern.

C. The officer should take a few seconds to look in and get the layout of the area.

D. The backup person should be on the same side of the hoseline as the nozzle operator.

107. An engine company officer was reviewing a recent strip mall fire with her firefighters, where the possibility of a backdraft explosion had seemed likely. She made the following statements about their options at such situations. She was correct in which statements?

1. When possible backdraft conditions are encountered, it is critical that all members keep clear of the storefront, especially plate glass windows and the parapet wall.
2. If roof ventilation is not possible, or will take too long, the unit will attempt an indirect attack using a fog stream injected through a large hole in a plate glass window.
3. If possible, the unit will break the window using the stream from the deck gun.
4. If it is impossible to break the glass from a distance, one member will have to break it from a safe area, and all other personnel will take cover behind the apparatus or parked cars.

A. all of the above

B. 1, 2, and 3 only

C. 1, 3, and 4 only

D. 2, 3, and 4 only

108. An engine officer made the following statements about potential backdrafts. Which statement is incorrect?

A. Backdrafts are most likely in tightly sealed buildings that have had a long time to get cooking.

B. Backdrafts are a result of an opening being made into a highly heated space where nearly all the oxygen has been consumed.

C. Backdraft explosions can be tremendously powerful blasts that level brick walls.

D. Backdrafts cannot occur at fires that already have flame venting to the outside.

109. A chief officer made the following statements regarding exposure protection. He was correct in all but which statement?

A. Fire units that are confronted by a massive body of fire that is beyond their ability to control initially should concentrate on protecting exposures.

B. Exposure protection should rely on a water curtain between the fire and the exposure.

C. Water applied to the surface of the exposure should be done using 2½-in. lines with a straight stream from a distance.

D. Select a location where you can alternate between hitting the fire and the exposure, and be prepared to go on the offensive when conditions permit.

110. Which statement concerning operations at sprinklered buildings is least correct?

A. If a serious fire is encountered in a sprinklered building, immediately call for assistance.

B. Fires in sprinklered buildings where the sprinklers are out of service are more dangerous than fires in buildings without sprinklers.

C. One reason that fires in sprinklered buildings are more dangerous is that exit travel distances are often doubled compared to an unsprinklered building.

D. If we find that the sprinklers are not working, we should stretch handlines and forget the sprinkler system.

111. When deciding whether to break a window for ventilation or to unlock and raise it, which factor is incorrectly stated?

A. If the smoke level in the room is so bad that you cannot stand up to manipulate the locks, break the glass.

B. Be aware of the status of the hoseline; if there is no water on the fire, hold off on breaking glass.

C. If there are reports of people trapped, breaking the glass is warranted.

D. Breaking double-hung windows provides twice the ventilation that opening them does.

112. What factors influence the choice of whether to use horizontal or vertical ventilation, or both?

1. size and location of the fire
2. construction of the building
3. available staffing
4. weather
5. availability of tools needed to perform each type

A. 1, 2, 3, and 4 only
B. 1, 3, and 5 only
C. 3 and 5 only
D. all of the above

113. A unit that arrives at a fire in new construction in a coastal community should expect to encounter hurricane glazing. To ventilate the windows in such a structure, the unit should use all but which technique?

A. Strike the glass very forcefully with any common tool.
B. Use an axe with short chopping strokes, nearly perpendicular to the pane.
C. Use a saw with a wood-cutting blade.
D. If these techniques are unsuccessful, prepare to cut through the wall.

114. All of the following but one are reasons why bowstring trusses are extremely dangerous. Which choice does not belong?

A. When a bowstring truss fails, it opens up a very large area nearly instantly.
B. Failure of one bowstring truss can cause a domino effect on other trusses.
C. When bowstring trusses fail, they often push out the end walls.
D. Bowstring trusses are difficult to detect from roof level.

115. The firefighters survival survey includes which of the following items? Select the most correct choices.

1. What is the occupancy?
2. Where are the occupants?
3. Where is the fire?
4. How do we get in?
5. How do we get out?
6. What is happening to the building?
7. What is the water supply?
8. What is the status of auxiliary appliances?

A. all of the above
B. 1, 2, 3, 4, 5, and 6
C. 1, 3, 4, 5, 6, and 7
D. 1, 2, 3, 5, 7, and 8

116. A RIT searching a smoke-filled second floor encounters a very large, unconscious firefighter who would have to be brought down a very narrow, winding stair. What statement below would be correct concerning the methods used to remove the victim firefighter?

A. The RIT should await the arrival of the rescue company with their mechanical advantage system before commencing this rescue.
B. A rope should be attached to the victim's wrists.
C. One rescuer drags the victim, head first, down the stairs.
D. The remaining members pull on the rope from below.

117. Select the incorrect statement regarding fires in buildings with metal deck roofs.

A. Metal deck roofs are usually supported on bar joists, which may be up to 60 ft long.
B. Bar joists may be spaced up to 6 ft apart.
C. Safety demands a ventilation hole of at least 4×4-ft be made on these roofs.
D. It may be necessary to cut sidewalls of these buildings to provide adequate ventilation.

118. An incident commander arrives to find a serious fire in an occupied senior citizens development. She knows from preplanning that it is built using 2×4-ft trusses with gusset plates. She directs all of the following actions, but one of them is ill-advised. Which one is incorrect?

A. ordering the attack to begin from a distance, using the reach of the stream
B. ordering a line operated into the truss loft by opening the exterior wall
C. ordering total window ventilation and maximum lighting
D. ordering a rapid advance into the structure once the fire has been knocked down

119. A fire officer training new members of his crew made the following statements concerning metal deck roof fires. The officer was most correct in which choice?

1. A metal deck roof fire can occur without any major involvement of the stock below the roof.
2. The metal deck acts like a gigantic frying pan when heated, distilling flammable gases from the tar.
3. These gases rise through the insulation, igniting above the roof.
4. The result is a self-accelerating reaction that can spread quite rapidly.
5. The reach and cooling power of 2½-in. lines are helpful on the roof.
6. It may be necessary to cut through the insulation to fully extinguish the roof fire.
7. Metal deck roof fires can occur in fully sprinklered buildings.

A. 1, 2, 3, and 6 only
B. 1, 2, 4, 5, and 6 only
C. 1, 2, 4, 6, and 7 only
D. all of the above

120. An engine company driver, having laid an inline supply from the most satisfactory hydrant, should spot her apparatus at which location at a two-story private dwelling fire?

A. just before the fire building
B. directly in front of the fire building
C. just past the fire building
D. at least two houses past the fire building

121. When discussing large multiple dwelling fires at a recent critique of an incident, a chief officer made the following statements. He was correct in which choice(s)?

1. Personnel must know where they are operating in the fire building.
2. A common frame of reference is needed so all personnel can accurately describe their location if they need help.
3. A building with several different wings is subdivided by letters, A, B, C, etc., starting with the right side.
4. It is helpful to have an aerial view of the structure to have an accurate idea of the spatial relationship between wings.
5. The roof team may have to communicate the layout of the structure to the IC.

A. 1, 2, 3, and 4 only
B. 1, 2, 4, and 5 only
C. 1, 2, 3, and 5 only
D. 1, 2, and 5 only

122. Fire that begins in a common crawl space under a row of garden apartments requires the proper tactics to control. Which choice below most correctly reflects the tactics to be used at these fires?

A. A large (2½-in.) hoseline must be stretched to cut off fire extension.
B. The first floor must be cut early to vent the fire area.
C. Horizontal ventilation is usually sufficient in these situations.
D. The cellar pipe, distributors, or the bent tip may be very helpful.

123. Which description below accurately describes the open framed stage of construction?

A. All structural supports are in place, but no walls, roofing, or flooring are in place.
B. All structural supports are in place, as well as walls, floors, and roof deck, but no windows, doors, or drywall are in place.
C. All structural supports are in place, as well as walls, floors, roof deck, windows, and doors, but no drywall is in place.
D. All structural supports are in place, and walls, floors, roof deck, windows, doors, and drywall are in place.

124. Which of the following choices most correctly describes a potential warning sign of impending backdraft?

1. heavy smoke, highly heated walls
2. late night or early morning fires with no visible smoke
3. smoke issuing under pressure
4. smoke being drawn back into the buildings

A. all of the above
B. 1, 2, and 3 only
C. 2, 3, and 4 only
D. 3 and 4 only

125. What factors make determining the fire floor and location so difficult at many high-rises?

1. the central air conditioning system that serves several floors
2. open (access or convenience) stairs
3. smoke travel in elevator and utility shafts
4. smoke travel in stairs via stack effect

A. 1 only
B. 1 and 2 only
C. 1, 2, and 3 only
D. all of the above

126. What is a rule of thumb for determining the number of personnel required to staff a stairwell support section at a high-rise fire?

A. one member per floor up to the fire floor
B. two members per floor up to the fire floor
C. one member per every two floors above the fire floor
D. one member per every two floors up to the fire floor

127. All of the following indicate the presence of a multiple dwelling except which choice?

A. a sign on the front that reads "Hotel Harriett Ann"
B. the presence of four or five gas or electric meters
C. the presence of a doorbell and mailbox
D. information received with alarm such as, "fire in Apartment 3E"

128. Immediately after knocking down a heavy fire condition in a ground floor apartment in a four-story building, you note fire still burning in the wall behind the toilet. Which of the following is the least correct action to take?

A. Order the hose stream directed up and down the pipe chase.
B. Have the members on the roof feel the soil pipes for heat.
C. Contact other members on the floor above and tell them about the problem.
D. When members on the roof report that one soil pipe is hot to the touch, order them to begin opening the roof around it.

129. Where heavy fire is evident in the cockloft of a large, multi-winged apartment building, what defensive measures should you prepare?

1. Position an elevating platform to cut off extension.
2. Cut trenches to isolate wings.
3. Position additional vent holes to draw fire to the trench.

A. all of the above
B. 1 only
C. 1 and 2 only
D. 2 and 3 only

130. Which choice correctly describes a difficulty present in many renovated multiple dwellings?

A. Drop ceilings create cocklofts on each floor that are interconnected with the vertical pipe chases, creating a maze for fire to travel in.
B. Fire originating outside these blind spaces often outflanks firefighting efforts.
C. The number of voids makes opening them a simple task.
D. Sprinklers in the public hallways are vital allies that can stop extension.

131. If you find yourself facing a blowtorch fire in the hall of a Class 1 multiple dwelling, you might employ a number of techniques to reduce the severe effects of the blaze. Which of the choices below would be most correct?

A. Push down the hallway with two handlines operating on straight stream.
B. Use a positive pressure fan to pressurize the ventilation stair.
C. Use a wind control device such as the K.O. curtain to block the exhaust window in the fire apartment.
D. Knock the fire down from the exterior using an aerial stream or a special nozzle from the floor below.

132. What is the most serious defect, from a fire spread standpoint, in a multistory private home?

A. lack of code requirements for interior finishes
B. lack of an enclosed stairway
C. large number of occupants
D. the small room size

133. Which statement regarding fire operations and exposure protection at private dwellings is least accurate?

A. Closely spaced exposures on two sides may require 250 gpm.
B. A master stream may be needed to stop fire extending down a row of closely spaced frame dwellings.
C. The hosebed and pumper-hydrant system must be capable of rapidly supplying this needed flow.
D. An inline stretch of large-diameter hose is the fastest way to apply these streams.

134. What statement below most accurately reflects the proper methods for cutting the roof of a peaked roof private dwelling?

A. An 8×8-ft hole is the recommended size of opening.
B. If operating from the basket of a platform, the quick cut is the preferred method.
C. The basket cut provides much less ventilation than the standard square hole.
D. The big advantage of the quick cut is that it only requires one set of nails to be pulled up.

135. When a firefighter is trapped or missing, the incident commander must take some specific actions, all of which are correctly listed below, except one. Which choice does not belong?

A. Commit all needed resources to the rescue effort, and forego any firefighting.

B. Clear radio channels of any unnecessary traffic, and establish clear lines of communication with the parties reporting the problem.

C. Transfer command of either firefighting or rescue operations to another officer.

D. Special-call any other needed resources, such as ALS (advanced life support) ambulances and additional ladder companies.

136. At a recent fire, an unconscious firefighter had to be brought down a flight of stairs to reach safety. Concerning this rescue, select the one correct statement below.

A. Whenever this situation is encountered, the blanket drag is the method to use.

B. One member starts down the stairs with the victim's head and shoulders, protecting this area from further damage.

C. The other member carries the legs and maintains the balance of the group.

D. Since gravity is helping under these circumstances, the smaller member should take the bottom position.

137. Which choice below most correctly reflects the hazards to firefighters in a new building under construction?

A. large quantities of flammable solids such as MAPP for steel cutting

B. shafts that are protected only by noncombustible materials

C. asbestos abatement operations

D. fire protection systems that have not kept pace with construction

138. What factors can cause smoke to travel counter to the normal diffusion pattern?

1. objects moving within shafts
2. central air conditioning (HVAC) systems
3. unusually cold weather during fire in an air-conditioned high-rise
4. fire in a very tall building

A. all of the above

B. 1 and 2 only

C. 1, 2, and 3

D. 2 only

139. An engine company arrives at a private home to find a strong odor of gas in front of the building and a hissing sound coming from the peck vent. They should take all but which of the following actions?

A. Notify the utility of a possible regulator failure.

B. Shut off the gas, at the curb cock if possible.

C. Pull the electric meter to remove sources of ignition.

D. Vent the building and search as needed.

140. A recent spate of persons reporting being shocked when they contacted metal streetlight poles prompted a captain to conduct a training session for her company. She made the following statements. Which choice reflects the most incorrect comment?

A. This situation is often caused by people illegally tapping into the streetlight's service to steal electricity.

B. The first action to take is to conduct a visual examination. Avoid touching the pole or wires without proper protection.

C. If wires are detected sticking out of the pole, a member wearing fire clothing should cut the illegal wires and wrap the exposed wires with electrical tape, tucking them back into the base of the pole to protect unsuspecting citizens.

D. Tape off the pole to further protect unsuspecting citizens, and notify the utility of the conditions.

141. Which building requires the greater collapse safety zone?

A. strip mall with a metal deck and bar joist roof

B. bowstring truss car dealership

C. supermarket with an inverted roof

D. bowling alley with hip rafters resting on the end walls of a bowstring truss roof

142. Which of the following statements concerning the behavior of steel when exposed to fire is least correct?

A. Steel is a brittle material that can fail suddenly.

B. When heated to about 1,000°F, a 100-ft. steel beam will expand almost 10 in.

C. When steel reaches 1,500°, it begins to fail.

D. When hot steel is hit with water, it freezes in shape and returns to its original length.

143. **The victims of a chemical attack can be triaged into four groups. Which group is least accurately described?**

A. Viable victims are able to respond to voice command or gentle touch.
B. Victims requiring advanced treatment respond only to painful stimuli.
C. Victims who do not respond even to painful stimuli are the highest priority for treatment in a mass casualty event.
D. The "walking well" are victims who show no signs or symptoms, but are fearful of having been exposed.

144. **Early one Sunday morning you arrive to find heavy fire blowing out of all the front windows of a two-story private house. Two cars are in the driveway, and neighbors tell you that a family of five live in the home, but have not been seen since the previous evening. What is the most correct action to take?**

A. Begin an aggressive interior search.
B. Begin defensive operations with emphasis on protecting exposures, since there is no possibility of survivors.
C. Begin an aggressive interior attack, knocking down as much fire as rapidly as possible, then searching for survivors.
D. Begin an aggressive interior attack, knocking down as much fire as rapidly as possible while simultaneously searching for survivors via VES of rear windows.

145. **What is the best method of reducing the life hazard before the fire?**

A. extinguishing the fire
B. using vent, enter, and search
C. reducing exposure hazards
D. reducing the life hazard before the fire through fire prevention

146. **Knowing the type of occupancy involved can tell an officer much about a given incident. Which of the following are key variables strongly tied to occupancy?**

1. potential life hazard
2. presence of wooden shake shingles
3. presence of large, open floor spaces or small rooms
4. presence of hazardous materials
5. exposure hazards
6. degree of fire loading
7. possible presence of truss construction (large, open floor areas)

A. 1, 2, 4, 6, and 7
B. 1, 3, 4, 6, and 7
C. 1, 4, 5, 6, and 7
D. all of the above

147. **Which class of construction is most resistant to collapse caused by fire?**

A. older type Class 1
B. newer type Class 1
C. newer type Class 4
D. older type Class 4

148. **Which of the following choices represents special firefighting problems due to their location?**

A. a cellar fire in a fireproof building
B. a top-floor fire in a fireproof apartment complex
C a fire on the second floor of a three-story warehouse
D. a fire on the second floor of a four-story hotel

149. On a hot, humid day, you are the incident commander at a fire in a two-story Class 5 frame house with a cellar. Your first attack hoseline has entered the structure and has knocked down a lot of fire. The crew returns to the street with their air cylinders depleted, and the officer approaches to inform you that they cannot find the main body of fire, and there is still fire free-burning in many void spaces. You consult your dispatcher and find that you are 20 minutes into the alarm. What is the most correct order to issue?

A. Have all units evacuate the building; it is time to change tactics.
B. Have only the original unit evacuate the building.
C. Have the original unit replace their cylinders and complete the overhaul.
D. Have a fresh unit enter the structure to complete the overhaul.

150. Prior to beginning the actual attack on a free-burning fire, which actions should the nozzle team take?

1. Attempt to locate and account for occupants.
2. Survey the structure for alternate escape routes and other fires in remote areas.
3. If you are approaching from below, take a quick look at the floor below the fire, and search it for occupants.
4. Begin ventilation above the fire.

A. 1 and 2 only
B. 1, 2, and 3 only
C. 2, 3, and 4 only
D. all of the above

151. Finding a heavy fire on the fifth floor of a six-story ordinary construction apartment house, the IC ordered the members of the second-arriving engine to assist the first engine in stretching their hoseline, even though it was clear that a second line would be required. The officer's actions were:

A. Incorrect. The second line is needed and should be stretched immediately.
B. Incorrect. The members of the second engine should immediately begin evacuating the occupants.
C. Correct. Two companies are needed to operate a hoseline under heavy fire conditions.
D. Correct. Get the first line in operation quickly, before stretching additional lines.

152. What factors determine the required diameter of the attack line?

1. time of day

2. area of the building (potential fire area)
3. size of the fire
4. occupancy

A. all of the above
B. 1 and 2 only
C. 2 and 3 only
D. 2, 3, and 4 only

153. The disadvantages of using a Bresnan distributor on a severe cellar fire include all but which choice?

A. It requires constant staffing to operate.
B. The operation area maybe untenable.
C. It applies water in a circular pattern in all directions at once.
D. The stream has limited reach, only 15 to 20 ft.

154. The advantage of 2½-in. hose over smaller diameter lines is incorrectly stated in which choice?

A. A larger flowing line has more reach than a smaller line at the same pressure.
B. A larger flowing line has greater penetration than a smaller line at the same pressure.
C. A larger line takes the same number of personnel to advance as a smaller line.
D. A larger line can be held by a single person.

155. A fully involved commercial occupancy 20 ft wide × 75 ft deep, common to many older strip malls, would require approximately how much flow to ensure rapid knockdown and extinguishment?

A. 100 gpm
B. 200 gpm
C. 350 gpm
D. 500 gpm

156. A senior firefighter in an engine company relayed her years of experience to a newer member. She made the following statements regarding hose stretching to upper floors. She was correct in all but which choice?

A. If a wide stairwell is encountered, it is best to stretch the line in the stairwell. The presence of such a well must be relayed to those stretching the hose to avoid stretching too much hose.

B. When stretching up a stairwell, do not stretch more than three lines in the well to avoid them getting tangled.

C. When stretching to upper floors via a wide well, one length of hose will usually reach up to the fifth floor. The length must be supported just below the coupling with a rope or hose strap.

D. If an elevator is installed in the well, it is best to use a rope to haul the line up the outside of the building if the fire is above the third floor.

157. Applying hydraulic knowledge to water supply problems allows firefighters to correctly supply hose streams. Which statement below is least correct regarding hydraulics?

A. By increasing the diameter of a hoseline, more water can be delivered at the same pump discharge pressure.

B. By increasing the length of a hoseline, less water can be delivered at the same pump discharge pressure.

C. By increasing the discharge rating of a pumper, less water can be delivered at the same pump discharge pressure.

D. By increasing the discharge rating of a pumper, more water can be delivered at a higher pump discharge pressure.

158. A senior firefighter instructing two newer members made the following statements about why fires in sprinklered buildings seem more smoky than fires in unsprinklered buildings. The member was correct in which choices?

1. Unsprinklered fires are more often free-burning fires.
2. Sprinklers do not allow complete combustion, creating CO.
3. Sprinklers cool the fire gases, making them more buoyant.
4. Sprinkler spray patterns push gases down like a fog nozzle.
5. Fires in unsprinklered buildings burn with less intensity.

A. 1, 2, 3, and 5 only

B. 1, 2, 3, and 4 only

C. 1, 2, 4, and 5 only

D. 1, 2, and 4 only

159. According to NFPA 14, standpipe systems equipped with booster pumps must have pressure control devices to limit the minimum and maximum pressure available. Which choice below accurately reflects the correct pressure limit at the top-floor hose outlet for that building?

A. in pre-1993 buildings, 100 psi minimum at the top floor outlet

B. in post-1993 buildings, 100 psi minimum for 2½-in. outlets

C. in post-1993 buildings, 175 psi maximum for 2½-in. outlets

D. in post-1993 buildings, 65 psi minimum for 1½-in. outlets

160. What is the most important item affecting ladder selection on the fireground?

A. nested length

B. strength

C. electrical conductivity

D. length

161. When planning the overhaul of a room that had just been knocked down by a hoseline, a ladder company officer made the following statements. He was least correct in which statement?

A. Plasterboard, or plaster on wire or wood lath, holds back a lot of fire.

B. Initial openings should be made near existing holes such as light fixtures.

C. If fire is found in any of these bays, the entire ceiling or wall must be opened.

D. Be especially alert for intersecting vertical openings where fire can climb up out of view.

162. What factors should not be included in the forcible-entry size-up?

1. occupancy
2. time of year
3. number of tools available
4. direction that the door opens
5. location of the fire, victims, and door to be used
6. type of door, jamb, and locks encountered

A. all of the above

B. 1, 4, 5, and 6 only

C. 2, 3, and 5 only

D. 2 and 3 only

163. When forcing an inward-opening door using the brute force method, what is the first way to attack the door?

A. using a sledgehammer to knock the door off its hinges
B. using a sledgehammer to knock a panel out of the center of the door
C. using a sledgehammer to breach a hole in the wall next to the lock
D. using a sledgehammer to drive a Halligan tool into the gap between the door and jamb

164. When encountering a padlocked security gate, it is usually better to cut the padlock than the gate itself. All but one of the following are correct reasons for this statement. Which choice is incorrect?

A. Cutting the gate requires a different saw blade than cutting the lock does.
B. Cutting the gate does not maximize access and ventilation.
C. It can take longer to cut the gate than to cut one or two padlocks.
D. Cutting the gate destroys it, which would not be justified by a minor fire or emergency.

165. What is a problem that is common to all operations utilizing any mechanical ventilation technique?

A. It requires additional personnel that may be more critically needed as part of the nozzle team.
B. Fresh air being drawn into the fire area can fan a smoldering fire into more serious proportions.
C. The ventilation we perform cannot necessarily be controlled as far as location and timing.
D. Ventilation causes a rapid buildup of toxic gases in remote areas prior to being cleared out.

166. What factors must you take into consideration when deciding to use positive pressure ventilation?

1. the location and extent of fire
2. the life hazard that may be affected by fire or venting
3. the availability of a search line
4. the degree of difficulty possible
5. debris that might be drawn into the exhaust
6. equipment available for the job, including the power supply
7. the locations of exhaust openings

A. 1, 2, and 3 only
B. 1, 2, 3, and 4 only
C. 1, 2, 6, and 7 only
D. all of the above

167. Which statement regarding a trench cut is least correct?

A. A trench cut is a defensive measure designed to slow fire spread to different areas of the fire building.
B. A trench cut should be the first action taken on the roof at serious fires in large buildings.
C. The best place to locate a trench is in the throat between wings of a building.
D. The trench should be not be opened until fire shows at the inspection holes on the fire side of the trench.

168. A quick search for live victims in the most likely areas they'd be, before the fire is under control, is a description of which phase?

A. search for life
B. search for fire
C. primary search
D. secondary search

169. Which choice below is least correct concerning secondary search techniques?

A. The secondary search may be combined with postcontrol overhauling.

B. Before moving objects such as draperies and chairs, thoroughly search them.

C. It is usually best to have different teams conduct the primary and secondary searches of an area.

D. Be sure to examine the perimeter, including rooftops and setbacks that people may jump to.

170. To perform a guide rope–assisted search, what items should members have on hand before entering the search area?

1. two-way radios
2. SCBA, PASS device, and flashlight for each member
3. search rope (400 ft or more of light line)
4. forcible-entry tools and TIC
5. large floodlight attached to the rope bag

A. 1 and 2 only

B. 1, 2, and 4 only

C. 2 only

D. all of the above

171. Question 5 of the firefighters survival survey is: "How do we get out when things go wrong?" Which choice is the most correct?

A. Most trapped firefighters become disoriented or lost prior to getting trapped.

B. Maintaining contact with a partner or a wall are the only safe ways to stay oriented

C. Firefighters must include a survey of the doors in their pre-entry size-up.

D. Be sure to watch for security gates and bars as you are making your escape.

172. Hazard awareness is critical to firefighter safety. Which of the following questions is not one of those outlined in the firefighters survival survey?

A. What is the occupancy?

B. What is the time of day?

C. Where is the fire?

D. How do we get in?

173. An officer instructing her members on the technique of the emergency body wrap with a personal rope cited the following statements as being critical to success. This officer was incorrect in which statement?

A. When descending, don't allow your hands to spread farther than about shoulder width apart.

B. Remove all slack from the rope just prior to exit.

C. Never attempt this maneuver without a turnout coat, gloves, and a second member belaying the member descending.

D. An NFPA-compliant one-person rope should be used for this escape.

174. Select the least correct statement regarding building construction from the choices below.

A. Virtually any type of building or occupancy can be found with either standard or lightweight construction.

B. The primary difference between standard and lightweight construction lies in the mass of the elements that carry the loads.

C. Older Class 2 or Class 5 buildings have relatively large members such as 2×10 joists or heavy, steel I-beams.

D. Using a standard wood beam, the maximum distance that can be spanned is about 25 ft.

175. What factors most correctly serve to indicate a bowstring truss?

1. classic humpback design
2. occupancy requiring large, open floor spaces without columns
3. occupancy used by bowling alleys, skating rinks, garages, etc.
4. a particularly deadly variety of bowstring truss has hip rafters that start on a truss and rest on an outer wall

A. 1 only

B. 1 and 2 only

C. 1, 2, and 3 only

D. all of the above

176. Many different styles of lightweight trusses are available, using short pieces of lumber joined together by various metal connectors. What danger do firefighting operations pose to all flat chord trusses?

A. The corrosion of the metal fasteners hastens collapse.
B. The small wooden elements burn through faster than large elements.
C. Hoseline operation tends to blow the connectors off.
D. If power saws are used to cut the deck, they may sever the top chord, precipitating collapse.

177. Select the most correct statement regarding fires and firefighting operations in private dwellings.

A. The majority of fires in private dwellings begin below the first floor.
B. The cellar often contains the heating plant, electric service, and water heater.
C. Quite a lot of cigarette smoking occurs in the below-grade areas.
D. Upper floors normally contain sleeping areas and must not be the primary target of VES attempts during daylight hours.

178. Roof ventilation on a private dwelling would be justified in all but which of the following cases?

A. fire in the cellar of a platform frame house
B. fire in an attic
C. fire in the cellar of a balloon frame house
D. fire that extends from the first floor via combustible siding to enter the attic via the eaves

179. Why is the furred-out space around an I-beam more of a threat for fire travel than a soil-pipe chase?

A. It is wider.
B. It goes to the cockloft.
C. It is harder to find.
D. all of the above

180. For a serious top-floor fire in a four-story apartment house, the roof team should take the following actions after they have vented over all vertical shafts. Which choice incorrectly describes the actions to take in this situation?

A. First, vent top floor windows from the roof.

B. Second, begin roof cutting.

C. If descent via fire escape is possible, venting and search from this area is preferred when fire is in proximity to the fire escape.

D. After the roof hole is cut, be sure to push down the ceilings below.

181. A chief officer discussing severe, wind-driven high-rise fires with several of his company officers made the following statements regarding these disastrous fires. He was least correct in which choice?

A. These fires have occurred in Class 1 residential buildings from 10 to more than 40 stories high.

B. These fires can occur on any floor of a building.

C. High wind (over 25 mph) is needed to fan these fires, which usually occur in winter.

D. Air flow through the building is the culprit, and it can create temperatures in excess of 2,500°F from standard residential fire loading.

182. Select the least correct statement regarding fires in garden apartments and/or townhouses.

A. Garden apartments are multiple dwellings.

B. Townhouses are single-family dwellings.

C. The term *condo* can accurately be used to describe either one.

D. Fire loading and compartmentation are similar in each.

183. Large garden apartment and townhouse complexes create serious fire problems. Select the least accurate statement regarding these structures from the choices below.

A. A 2½-in. handline with a break-apart nozzle attached to 100 ft. of 1¾-in, is a key tool for dealing with those apartments that are within the reach of a preconnect.

B. The 2½-in. line can be crucial if heavy fire is present and extending to exposures.

C. The 2½-in. line serves to decrease friction loss, while the 1¾-in. line allows making the many bends and turns inside the structure.

D. If a nozzle team has to back out due to heavy fire, disconnect the 1¾-in. line and immediately place the stream in operation.

184. Heavy fires involving a garden apartment or townhouse in the occupied stage are best fought with which hose streams?

A. a preconnected master stream

B. a 1¾- or 2-in. line through the front door

C. two 1¾-in. lines operating side by side

D. a 2½-in. handline with a large solid tip

185. A chief officer conducting a preplanning session at an older row of stores built with a standard flat roof gave the following instructions to the fire companies assembled at the scene. Which instruction does not correctly reflect a recommended tactic for these situations?

A. The roof team should attempt to cut an 8×8-ft vent hole over the main body of fire.

B. If further ventilation is needed, continue cutting additional vent holes.

C. If the vent holes do not stop fire spread, cut a trench to limit fire spread to other stores.

D. For a trench to be effective, it must be cut from outside wall to outside wall (or to a fire wall) and be subdivided every 4 ft.

186. A serious fire exists in a newer taxpayer built with a metal deck on bar joist roof. The roof team reports that the roof is sagging over the fire store. Which orders below most accurately portray proper tactics?

A. Order the roof evacuated.

B. Order roof forces to back away 20 ft from the danger area, then resume operating.

C. Pull interior forces back out of the involved area, then operate in exposed stores until the main body of fire is darkened down and the steel has cooled.

D. Pull interior forces back out of the entire building until the main body of fire is darkened down and the steel has cooled.

187. What items are included in the high-rise strategic plan?

1. Determine the general fire area.
2. Verify the location of the fire before committing handlines.
3. Take control of evacuation.
4. Gain control of building systems.
5. Confine and extinguish the fire.

A. all of the above
B. 1, 2, and 3 only
C. 2, 3, and 4 only
D. 2, 3, 4, and 5 only

188. What feature of high-rise office building construction listed below is not responsible for fire and smoke spread from floor to floor?

A. HVAC systems that serve more than one floor
B. curtain wall construction
C. access stairs
D. scissor stairs

189. All of the following are reasons why fires in buildings under construction, renovation, and demolition account for a large number of serious fires as well as firefighter injuries, except for one statement. Which choice does not belong?

A. numerous sources of ignition
B. large amounts of concealed combustibles
C. lack of adequate fire protection features
D. unlimited air supply if windows are not intact

190. Preplanning of large construction sites is vital to fire suppression efforts. What are the key items to document when visiting a structure?

1. large, undivided floor areas
2. enclosed vertical openings
3. limited access to upper floors
4. storage of flammable gases in temporarily enclosed areas
5. fire protection equipment not keeping pace with construction
6. buildings fully occupied

A. all of the above
B. 2, 3, and 5 only
C. 1, 2, 3, and 6 only
D. 1, 3, 4, and 5 only

191. Natural gas has a mercaptan compound added to it to make it recognizable. What statement below is most correct in the event of a large above-ground leak?

A. The odorant can escape, making the natural gas undetectable at the leak source.
B. The odorant can create a severe fire hazard at the leak.
C. The odorant can settle out upwind, making people believe they are in danger when in fact no gas is present.
D. The odorant is added in tiny quantities, only ¼ lb per million cubic ft of gas, making it possible to detect the leak by smell before it reaches the flammable range.

192. What is the only way to prevent a BLEVE of a fire-exposed LPG cylinder?

A. Extinguish the fire.
B. Activate the relief valve on the container.
C. Cool the lower liquid propane portion of the cylinder.
D. Remove the heat from the upper vapor space of the tank shell.

193. Select the most correct statement regarding the electrical system.

A. An open circuit does not permit electricity to flow due to an open switch or break in a wire.

B. A short circuit does not allow electricity to flow due to a break in a wire.

C. The utility ground is a wire installed by the utility to conduct electricity to the shortest path.

D. A house ground is a condition where the entire house is electrically grounded to the utility system.

194. A captain of a very busy ladder company was conducting a class for the members of the unit on electrical hazards in firefighting and made the following comments regarding possible injurious effects from electrical contact. Which statement is most accurate?

A. Electrocution is unaffected by several variables, including the amount of current passing through the body, the condition of the skin, and the duration of the contact.

B. The path the current takes through the body is an important factor in determining the survival of a person contacting electricity.

C. Current flow that traverses the chest is severe since it affects the brain and other vital organs.

D. Current that enters through the head and exits through the feet damages the brain and nervous system; current flow the opposite direction is not a problem.

195. A newly promoted lieutenant, wishing to impress her new unit, gave a drill regarding a recent transformer fire. She made the following statements. Which choice is least correct?

A. Assume a defensive position, with the apparatus downwind and downhill from the transformer if possible.

B. Secure a water supply, but do not flow any water until assured by utility personnel that the power is off except to protect exposures.

C. Once power has been removed, use dry chemical or foam streams to extinguish the burning oil.

D. Do not begin overhaul until advised by the utility about the presence of PCBs; consider using hazardous materials personnel for overhauling.

196. An engine company arrived first-due at an automobile versus pole accident that caused the electrical wires on the pole to break and fall on the car. The acting officer in the right front seat gave the following commands. Which one could have caused a tragic outcome if complied with as stated?

A. positioned the apparatus to detour traffic while requesting law enforcement assistance for crowd and traffic control

B. created a safety zone around the downed wires that was at least 10 ft in all directions from the wire

C. ordered a handline with a fog nozzle stretched but kept at least 25 ft from any electrical equipment involved

D. directed the occupants of the car to remain inside and clear of any metal parts until the power could be removed

197. Regarding structural collapse of high-rise office buildings, which statement is most accurate?

A. Buildings of modern Class 1 construction show the greatest resistance to collapse.

B. Modern Class 1 construction does not allow enough time for a long stair climb, evacuation, fire control efforts, and a long climb down before collapse.

C. Buildings that were constructed with an unprotected steel skeleton are most resistant to collapse from fire.

D. The exception to choice C is precast concrete construction after occupancy.

198. In which type building, framed or unframed, is a collapse likely to be more serious?

A. an unframed building, because it is built of all combustible materials

B. a framed building, because it is built of all combustible materials

C. an unframed building, because it depends on bearing walls

D. a framed building, because it depends on bearing walls

199. An intelligence officer briefing fire department leaders on some of the criteria that should be considered in evaluating potential terrorist threats was reported to have made the following statements. The officer must have been misquoted in which statement, if his briefing was to be completely correct?

A. Easily recognizable landmark buildings are prime targets.

B. Financial institutions and corporate headquarters are also threatened.

C. Any response involving a report of a chemical odor or a report of "a person is unconscious for an unknown reason" should be approached with caution.

D. All locations where an explosion has occurred (including reported transformer explosions) should be treated as potential terrorist events.

200. The role of the fire department at an RDD explosion is very important. What is one of the most important actions for first-arriving units to take?

A. Extinguish the fire in the debris, which sends more contaminated smoke skyward.

B. Use a solid stream to knock down radioactive plumes.

C. Keep all people out of the runoff water.

D. Use chemical detectors to survey the area.

PART IV

ANSWERS

ANSWERS

1 — GENERAL PRINCIPLES OF FIREFIGHTING

1. **A.** (p. 7)
2. **C.** (pp. 7–8)
3. **A.** (pp. 8) When several potential victims are present in different areas, putting the hoseline in operation to protect the means of egress is critical.
4. **C.** (p. 10) Remove those in greatest danger first, if they are savable.
5. **B.** (pp. 8–9)
6. **C.** (p. 11)
7. **A.** (p. 12)
8. **B.** (p. 7) Fire spread to a windowless Class 1, 3, or 4 building across a 20–ft alley will be very slow.
9. **D.** (p. 9–10) The size-up only indicates that the front is involved. Survivors in the rear should not be written off too early.
10. **D.** (pp. 11) Having four people permits both simultaneous actions at a serious situation with a person confirmed trapped. There might be others trapped as well.

2 — SIZE-UP

1. **C.** (p. 15)

2. **C.** (p. 15)
3. **D.** (p. 15)
4. **A.** (p. 15) All of the remaining items may or may not be true and must be verified on arrival.
5. Verify in the field, in your town.
6. **D.** (p. 16)
7. (pp. 10–16)

   C: construction
   O: occupancy
   A: apparatus and personnel
   L: life hazard
   W: water supply
   A: auxiliary appliances
   S: street conditions
   W: weather
   E: exposures
   A: area
   L: location and extent of fire
   T: time
   H: height

8. **C.** (p. 16)
9. **D.** (p. 17)
10. **C.** (pp. 17–18)
11. **B.** (p. 18)
12. **C.** (p. 23–24)
13. **A.** (pp. 24–27)
14. **D.** (pp. 20–23)
15. **C.** (pp. 27–28)
16. **D.** (pp. 29–30) This fire is not below grade, not out of reach of ladders, and *not* windowless, nor is it a top floor fire in a Class 3 or 5 building.

17 **A.** (pp. 29–30)

18. **C.** (pp. 31)
19. **D.** (pp. 23)
20. **B.** (p. 35)
21. (p. 35)

    fatigue firefighters
    slower operations/mechanical failures
    drive attack crews off the floor/past defensive measures

22. **A.** (pp. 36)
23. **D.** (pp. 19–23)
24. **D.** (pp. 23)
25. **C.** (p. 24)

3 — ENGINE COMPANY OPERATIONS

1. **D.** (pp. 39)
2. **D.** (p. 40)
3. **A.** (pp. 40)
4. **D.** (pp. 41–42)
5. **C.** (pp. 45)
6. **B.** (pp. 41)
7. **C.** (p. 42)
8. **C.** (p. 43)
9. **A.** (p. 42)
10. **A.** (p. 55)

11. **A.** (pp. 52–53)
12. **C.** (p. 53)
13. **D.** (p. 56)
14. **B.** (pp. 56–57)
15. **D.** (p. 58)
16. **C.** (p. 58)

4 — HOSELINE SELECTION, STRETCHING, AND PLACEMENT

1. **B.** (pp. 61) height and area
2. **D.** (p. 66)
3. **B.** (pp. 65–67)
4. **A.** (pp. 63–64)
5. **B.** (p. 65)
6. **C.** (pp. 67)
7. **B.** (pp. 67–69)
8. **C.** (pp. 69) 600 ft: 200 ft + 200 ft (length and width), plus two lengths for the two flights of stairs, plus 100 ft from hydrant to entrance
9. **A.** (p. 72) At least one length for the fire apartment and one length up the five flights via the open stairwell.
10. **B.** (pp. 73–74) The first line must go to the interior stairs on the fire floor, just as a line coming up the staircase would.
11. **A.** (p. 81)
12. **B.** (p. 81)
13. **C.** (pp. 82–83)
14. **B.** (p. 83)
15. **D.** (pp. 90–91)
16. **C.** (pp. 63–64)
17. **C.** (p. 65)
18. **A.** (pp. 66–67)
19. **C.** (p. 67)
20. **B.** (pp. 70)
21. **C.** (pp. 69–75) just below the coupling
22. **B.** (pp. 75)

5 — WATER SUPPLY

1. **A.** (p. 96)
2. **D.** (pp. 96)
3. **B.** (p. 97)
4. **D.** (pp. 98)
5. **B.** (pp. 103)
6. **C.** (pp. 100)
7. **D.** (p. 101–102)
8. **B.** (p. 104)
9. **A.** (p. 107–111)
10. **B.** (pp. 115)
11. **A.** (p. 118)
12. **C.** (p. 96)
13. **C.** (p. 367–368)
14. **D.** (p. 98)
15. **C.** (pp. 100)

16. **C.** (pp. 100)
17. **B.** (pp. 102–104)
18. **B.** (pp. 101–114)
19. **B.** (pp. 101–114)

6 — SPRINKLER SYSTEMS AND STANDPIPE OPERATIONS

1. **D.** (p. 119)
2. **A.** (p. 119)
3. **D.** (p. 119)
4. **B.** (p. 121) (1) A combination of hoselines and sprinklers may be holding the fire in check, or (2) a second fire may be burning in a remote area.
5. **D.** (p. 121)
6. **A.** (p. 121)
7. **C.** (p. 121) It will cause the sprinkler operating pressure and discharge to drop.
8. **D.** (pp. 121–122) Stretch one of the first hoselines to supply the Siamese and back it up with a second line.
9. **B.** (p. 122)
10. **D.** (p. 122)
11. **B.** (p. 122)
12. **C.** (p. 122)
13. **B.** (pp. 123)
14. **A.** (pp. 123)
15. **C.** (pp. 123) sprinklers that begin to operate *after* fire department operations have begun
16. **C.** (pp. 123) in buildings that are only partially sprinklered, where the fire begins in an unsprinklered area
17. **D.** (p. 123) They negate venting due to their spray pattern.
18. **B.** (pp. 126–126)
19. **B.** (p. 126)
20. **D.** (p. 127)
21. **A.** (pp. 127–128)
22. **D.** (p. 127)
23. **D.** (pp. 128–129) Deluge systems utilize open heads throughout the area. Pre-action systems have closed heads, and only those over the fire discharge.
24. **A.** (pp. 130) Note: B: deluge system, C: dry pipe system, D: wet pipe system
25. **A.** (p. 132) Note: B, C, and D are all useful outward signs if A has not been done.
26. **D.** (p. 138)
27. **B.** (pp. 141)
28. **C.** (pp. 141–142) Run two supply lines of the largest hose available to each Siamese, and supply lines into the first-floor hose outlets.
29. **C.** (pp. 143–144)
30. **C.** (p. 145)
31. **B.** (pp. 152–154) 14- or 18-in. pipe wrench
32. **C.** (p. 154) seventh floor or below

walk, eighth or above ride partway

33. **D.** (pp. 142)

7 — LADDER COMPANY OPERATIONS

1. **D.** (p. 158)
2. **C.** (p. 158)
3. **D.** (p. 159)
4. **A.** (pp. 158–159)
5. **C.** (p. 159) others in nearby positions
6. **D.** (p. 158–159) (more than 70% of civilian fire deaths)
7. **B.** (pp. 159)
8. **D.** (p. 161)
9. **A.** (p. 162) The bedroom has the highest life hazard regardless of the time of day.
10. **B.** (pp. 165–168)
11. **D.** (p. 166)

12 **D.** (p. 167)

13. **B.** (p. 168)
14. **C.** (p. 168)
15. **C.** (p. 168)
16. **B.** (p. 172)
17. **C.** (p. 172)
18. **D.** (p. 174) Raise before retracting.
19. **D.** (pp. 176) the smaller the scrub area
20. **B.** (p. 177)
21. **D.** (p. 177)
22. **D.** (p. 180)
23. **D.** (pp. 181–183) sight, hearing, touch, smell, and common sense
24. **A.** (p. 183)
25. **D.** (pp. 185)
26. **B.** (pp. 186) Do not remove a lintel supporting a brick wall, or collapse may result.
27. **C.** (p. 186–187) Note: Tarps only protect against water and dust, not fire.
28. **A.** (p. 187)
29. **D.** (p. 186–187)
30. **B.** (p. 188)
31. **D.** (p. 165)
32. **A.** (p. 168)
33. **C.** (pp. 179–184)

8 — FORCIBLE ENTRY

1. **D.** (pp. 192)
2. **D.** (pp. 189)
3. **C.** (pp. 190–191)
4. **B.** (p. 193)
5. **A.** (p. 194)
6. **B.** (p. 195)
7. **D.** (pp. 195–197) All other choices destroy the integrity of the door.

8 **D.** (p. 197)

9. **A.** (pp. 197)
10. **B.** (p. 199)
11. **D.** (p. 200)
12. **B.** (p. 201)
13. **A.** (p. 205)
14. **C.** (p. 208–209) Note: The cylinder of a fox lock is recessed inside the door.
15. **C.** (p. 209)
16. **D.** (pp. 205)
17. **B.** (p. 210)
18. **D.** (p. 210)
19. **C.** (p. 214)
20. **D.** (p. 215)
21. **B.** (pp. 215–216)
22. **D.** (p. 217)
23. **C.** (p. 217)
24. **A.** (pp. 217)
25. **A.** (p. 218–220)
26. **D.** (p. 218)
27. **A.** (p. 219) Note: same amount of cutting.
28. **A.** (p. 220) Warn all members, especially those on the perimeter.
29. **D.** (p. 221–222) A chain saw would be used to cut the inner braces, as the circular saw does not cut deep enough to reach them.
30. **A.** (p. 228) B: You must be inside to operate them, so through-the-lock method is not possible. C: The key from the previous lock in the sequence is required, not just any lock.

9 — VENTILATION

1. **B.** (p. 230)
2. **C.** (pp. 231)
3. **C.** (p. 231)
4. **B.** (p. 231)
5. **C.** (p. 231)
6. **C.** (p. 234)
7. **D.** (p. 238) especially wind
8. **C.** (p. 238–239)
9. **B.** (pp. 237–238)
10. **D.** (p. 239)
11. **A.** (p. 240) Note: For choice 4, mechanical ventilation may be useful, but given the description of the blaze, it would not normally be a prerequisite.
12. **B.** (p. 240)
13. **A.** (pp. 240)
14. **C.** (p. 241) intake screen
15. **D.** (p. 243–244)
16. **C.** (p. 244)
17. **B.** (p. 244–245)
18. **A.** (p. 247, top)
19. **B.** (p. 248, bottom)

20. **C.** (p. 248)
21. **A.** (pp. 248)
22. **C.** (p. 250)
23. **D.** (p. 251)
24. **D.** (p. 252) Note: Cut F is not strictly required, but should be made anyway, to provide flexibility in the event of wind shifts.
25. **C.** (p. 25–252)
26. **B.** (p. 248) Note: There are no trusses in a standard roof.
27. **D.** (pp. 254)
28. **B.** (pp. 254)
29. **A.** (pp. 255) Two cocklofts create multiple places for the fire to hide.
30. **C.** (p. 256–257)
31. **C.** (pp. 257–258)
32. **D.** (pp. 258–260) inspection holes, not vent holes
33. **A.** (pp. 258–260) should include pulling the trench
34. **B.** (pp. 235–236) You must vent early, while the attack crew is in a safe position or delay venting until after the hose stream has thoroughly cooled the area.
35. **C.** (p. 261)
36. **B.** (p. 261–262)
37. **C.** (pp. 262)
38. **B.** (p. 266) particularly trusses

10 — SEARCH AND RESCUE

1. **B.** (p. 268)
2. **C.** (p. 268)
3. **C.** (p. 268)
4. **D.** (p. 268)
5. **C.** (pp. 270–272))
6. **C.** (pp. 272–273)
7. **B.** (p. 274)
8. **A.** (p. 275)
9. **D.** (p. 276)
10. **D.** (pp. 277)
11. **D.** (p. 277)
12. **D.** (p. 280)
13. **B.** (p. 280)
14. **C.** (p. 280) minimum two members per apartment
15. **D.** (p. 274–277)
16. **B.** (p. 278–288) Search where the objects will land.
17. **D.** (pp. 280–281)
18. **B.** (pp. 282) In this use, the rope is intended to provide a means of avoiding disorientation. One member may hold it while others work nearby without being tied off, if smoke and visibility permits.
19. **C.** (p. 274)
20. **A.** (pp. 269) each member

11 — FIREFIGHTER SURVIVAL

1. **C.** (p. 285)
2. **A.** (p. 287)
3 **B.** (pp. 286–287)
4. **A.** (p. 287)
5. **A.** (p. 287–288)
6. **D.** (p. 288)
7. **A.** (pp. 288–289)
8. **D.** (pp. 289–291)
9. **C.** (p. 295) Lie with face at floor and lights off *momentarily*. Do not walk, crawl with your light on.
10. **C.** (pp. 295) A belay is only used in training.
11. **D.** (p. 296) Post second copy on dash, and give third copy to command post.
12. **C.** (p. 298) Note: A: In the scenario the member has already been reported as missing. B: Good, but might not be needed; information is always required. D: The member may be buried, her PASS device may not be armed, or she might be okay and just unaccounted for.
13. **B.** (p. 300)
14. **D.** (pp. 301)
15. **B.** (p. 302)
16. **D.** (p. 303)
17. **C.** (p. 303)
18. **D.** (pp. 303)
19. **D.** (p. 304)
20. **B.** (pp. 305) Note: Until after fire is under control.
21. **D.** (p. 306) The spare bottle stays on the rig unless needed.
22. **B.** (p. 306) Note: A: one for each two-person team. C: one for each two-person team. D: mask for each trapped firefighter
23. **A.** (p. 306) Note: Lifesaving rope (two-person), not a personal rope (one-person) is needed, and you bring this with you, not try to locate one at the scene.
24. **D.** (p. 306)
25. **A.** (p. 315–316) Note: It should be an advanced life support (ALS) unit, not BLS.
26. **C.** (pp. 307–308) This is a manual operation, but the member is unconscious.
27. **D.** (p. 308) All are drawbacks; the question asked which one is *not*. Pay attention to the wording of the question!
28. **B.** (pp. 309–310) A definite time limit is set for each member to operate, based on their SCBA duration.
29. **D.** (pp. 309–312)
30. **C.** (pp. 315)
31. **D.** (p. 316) There is no guarantee the firefighter is conscious or able to assist.
32. **C.** (pp. 317–318)
33. **C.** (pp. 317)

34. **B.** (pp. 318) Note: The exception may be for a lifesaving purpose, where the member is trapped and cannot be moved, and where help is coming with additional air supplies.

35. **B.** (pp. 319–320)

36. **C.** (pp. 322–323)

37. **A.** (pp. 324)

38. **A.** (pp. 323–326)

39. **C.** (pp. 327–329) *same* extremities

40. **C.** (p. 328–329)

12 — OPERATIONS IN LIGHTWEIGHT BUILDINGS

1. **C.** (p. 333)
2. **C.** (p. 333)
3. **D.** (p. 334)
4. **C.** (p. 335)
5. **D.** (pp. 335)
6. **B.** (pp. 335–336)
7. **B.** (pp. 336)
8. **A.** (p. 336)
9. **A.** (p. 337)
10. **C.** (pp. 337)
11. **D.** (p. 339)
12. **B.** (p. 340)
13. **D.** (p. 340)
14. **A.** (p. 341)
15. **C.** (pp. 342)
16. **C.** (p. 343)
17. **C.** (pp. 343)
18. **B.** (pp. 343–345)
19. **C.** (p. 345)
20. **D.** (pp. 345–346)
21. **B.** (p. 346)
22. **D.** (pp. 347)
23. **D.** (p. 347)
24. **D.** (pp. 347–348)
25. **A.** (p. 348)

13 — BELOW-GRADE FIRES: BASEMENTS AND SUBBASEMENTS, CELLARS AND SUBCELLARS, AND CRAWL SPACES

1. **D.** (pp. 367–368)
2. **D.** (pp. 370) Note: The area over the fire must also be vented.
3. **A.** (pp. 370–371)
4. **A.** (p. 372)
5. **A.** (p. 372)
6. **D.** (p. 373)
7. **B.** (p. 374)
8. **C.** (p. 374) 200 ft of 2½-in. hose (twice the depth = 150 ft. plus one length for stairs)

9. **D.** (p. 374) A wide fog will produce tremendous steam conditions, burning members, and could push fire around behind the line.

10. **A.** (p. 374)

11. **D.** (pp. 373–374)

12. **C.** (pp. 374)

13. **A.** (p. 375)

14 — PRIVATE DWELLINGS

1. **C.** (p. 379)
2. **D.** (p. 379)
3. **D.** (p. 379)
4. **A.** (p. 386)
5. **C.** (p. 380)
6. **D.** (p. 380)
7. **C.** (pp. 381) three lines of 1½- to 2-in. diameter
8. **B.** (p. 361)
9. **D.** (p. 381–382)
10. **C.** (p. 381) just over 2½ minutes (at 175–180 gpm)
11. **B.** (p. 383)
12. **B.** (pp. 382–383)
13. **C.** (p. 385–386)
14. **A.** (p. 386)
15. **D.** (p. 387)
16. **C.** (p. 386) Note: D: After the hole is cut, the hole must be pulled open, then the ceiling below pushed down.
17. **B.** (p. 389)

Ensure the stability of the area before entering.

Always have two means of escape.

Plan the cut and inform other members of its layout.

Keep the wind at your back.

Cut adjacent to joists.

Never step on the cut.

Don't cut the roof supports.

18. **A.** (p. 389)
19.

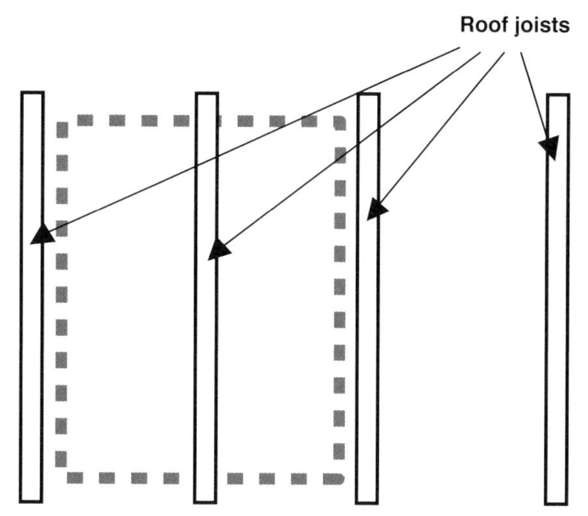

15 — MULTIPLE DWELLINGS

1. **C.** (p. 391)
2. **D.** (p. 391) due to the varying sleep patterns of a large number of occupants (e.g. shift workers, night workers)

3. **B.** (pp. 391–392)

4. **A.** (p. 392)

5. **A.** (p. 393–394)

6. **D.** (p. 394)

7. **D.** (p. 394–395) There are often no visible indications of where this space is located. It is easier to find a toilet or sink in heavy smoke, and open up behind it.

8. **C.** (p. 395) Note: A: Not all, only those in great danger; protect the rest in place. B: Same thing, don't waste time and precious personnel when putting the fire out solves the problem.

9. **D.** (p. 396) Note: The member venting the window on the stair half-landing above the fire floor needs an axe or heavy tool. These windows are often wire glass.

10. **D.** (pp. 397)

11. **B.** (p. 397)

12. **C.** (p. 397)

13. **D.** (p. 398)

14. **B.** (p. 400)

15. **A.** (p. 400)

16. **C.** (p. 401)

17. **C.** (p. 401)

18. **D.** (p. 401) via an alternate route, e.g., by rope on the exterior or via another staircase

19. **A.** (pp. 403) The worst-case scenario occurs in O-shaped buildings, allowing fire spread in two directions at once.

20. **D.** (p. 404)

21. **B.** (pp. 405)

22. **C.** (p. 406)

23. **C.** (pp. 406–407)

24. **A.** (pp. 407)

25. **B.** (pp. 408)

26. **A.** (pp. 409)

27. **D.** (p. 410)

28. **B.** (pp. 412–413)

29. **C.** (p. 412)

30. **B.** (pp. 416–417)

31. **D.** (pp. 416)

32 **C.** (p. 417)

33. **D.** (p. 417)

16 — GARDEN APARTMENT AND TOWNHOUSE FIRES

1. **D.** (p. 419)

2. **B.** (p. 419) Compartmentation within the apartment; garden apartments in particular are likely to have large open cocklofts over several apartments.

3. **A.** (pp. 419)

4. **C.** (pp. 419–420)

5. **A.** (p. 420–421)

6. **D.** (p. 421–422)

7. **C.** (pp. 422)
8. **C.** (p. 422)
9. **D.** (p. 423)
10. **B.** (p. 424)
11. **A.** (p. 425)
12. **B.** (pp. 426)
13. **D.** (p. 427)
14. **C.** (p. 427) Paint the designation
15 **C.** (p. 427)
16. **D.** (pp. 428)
17. **C.** (pp. 424)
18. **D.** (p. 424)

17 — STORE FIRES —TAXPAYERS AND STRIP MALLS

1. **C.** (p. 429)
2. **B.** (p. 429)
3. **C.** (p. 430)
4. **B.** (p. 429)
5. **A.** (p. 430)
6. **C.** (pp. 430)
7. **A.** (p. 431)
8. **C.** (p. 431)
9. **C.** (p. 432–433)
10. **B.** (pp. 431–434)
11. **D.** (pp. 434–436)
12. **B.** (pp. 434–435)
13. **B.** (pp. 435)
14. **A.** (pp. 439)
15. **B.** (p. 436)
16. **C.** (p. 437)
17. **B.** (p. 437)
18. **A.** (p. 437)
19. **B.** (p. 438) Note: A: For heavy fire, use a master stream in fire store and 1¾-in. or 2-in. lines in exposures. C: More than one line is needed per store. D: From sidewalk level
20. **D.** (p. 441)
21. **D.** (p. 442)
22. **D.** (p. 444)
23. **A.** (p. 445)
24. **A.** (p. 444)
25. **D.** (p. 445)

18 — HIGH-RISE OFFICE BUILDINGS

1. **C.** (p. 448)
2. **C.** (p. 448)
3. **D.** (p. 448–449)
4. **D.** (p. 449)
5. **A.** (p. 449)
6. **D.** (p. 449–450)
7. **B.** (pp. 449–450) Evacuation is not normally required *initially*, but is

essential for large fires.

8. **A.** (p. 449–450)
9. **C.** (pp. 450)
10. **A.** (p. 451)
11. **B.** (pp. 451, 460)
12. **A.** (pp. 456–457)
13. **B.** (p. 457)
14. **D.** (pp. 460)
15. **B.** (pp. 460)
16. **C.** (p. 460)
17. **B.** (p. 461)
18. **D.** (p. 461)
19. **D.** (p. 461)
20. **B.** (pp. 461–462)
21. **A.** (pp. 461–465)
22. **A.** (p. 463)
23. **D.** (pp. 462–465)
24. **C.** (pp. 465)
25. **C.** (p. 466)
26. **B.** (pp. 466–467)
27. **D.** (p. 467)
28. **D.** (pp. 470–471)
29. **C.** (pp. 473–475)
30. **B.** (pp. 473)
31. **D.** (p. 473)
32. **C.** (pp. 474–475)
33. **A.** (p. 475)
34. **D.** (pp. 475)
35. **A.** (pp. 478)
36. **B.** (pp. 477–478) Operations only requires floor plans for the fire floor and floor directly above, not all floors. These should be left at command for possible use by the SAE (search and evacuation) post.
37. **A.** (pp. 478–479) Coordinating multiple units in a staircase is the job of an attack chief. The operations officer coordinates multiple staircases.
38. **D.** (p. 478–480)
39. **D.** (pp. 480)
40. **B.** (p. 480) Verify searches above the fire area; operations controls the fire floor and one floor above.

19—BUILDINGS UNDER CONSTRUCTION, RENOVATION, AND DEMOLITION

1. **D.** (p. 481)
2. **A.** (p. 481)
3. **C.** (p. 482)
4. **D.** (pp. 482–485)
5. **A.** (p. 482)
6. **B.** (p. 482–483)
7. **B.** (pp. 482–490)
8. **B.** (485)
9. **D.** (pp. 486)

10. **C.** (p. 487)
11. **B.** (p. 487)
12. **A.** (pp. 487–488)
13. **C.** (pp. 488)
14. **C.** (p. 489)
15. **D.** (p. 489)
16. **B.** (p. 490
17. **B.** (p. 491)
18. **A.** (pp. 491)
19. **D.** (pp. 493)
20. **D.** (pp. 492–493)

20 — FIRE-RELATED EMERGENCIES: INCINERATORS, OIL BURNERS, AND GAS LEAKS

1. **B.** (p. 495)
2. **A.** (p. 495)
3. **D.** (pp. 495–498)
4. **D.** (pp. 495) The odorant settles out downwind.
5. **A.** (p. 503)
6. **C.** (p. 496)
7. **C.** (p. 496)
8. **A.** (p. 498)
9. **D.** (p. 498)
10. **B.** (p. 498)
11. **C.** (p. 498)
12. **D.** (p. 499)
13. **A.** (p. 500)
14. **C.** (p. 500)
15. **B.** (p. 500–501)
16. **B.** (p. 502)
17. **A.** (pp. 503)
18. **D.** (p. 502–503)
19. **D.** (p. 504)
20. **D.** (p. 504) High-pressure leaks should not be stopped by plugging; instead, close valves.
21. **B.** (p. 507)
22. **C.** (p. 508) from 100 psi to 190 psi
23. **D.** (p. 508)
24. **D.** (p. 509)
25. **B.** (p. 509)
26. **C.** (p. 507)
27. **C.** (p. 511)
28. **B.** (p. 512)
29. **D.** (pp. 513) Turn off the emergency switch and oil tank valve, and shut the burner door.
30. **C.** (pp. 518)
31. **C.** (pp. 520)
32. **A.** (p. 520)
33. **C.** (p. 521)

21 — ELECTRICAL FIRES AND EMERGENCIES

1. **C.** (pp. 523)
2. **C.** (p. 523–524)
3. **A.** (p. 524)
4. **A.** (pp. 524)
5. **C.** (p. 524–525)
6. **D.** (p. 524–525)
7. **B.** (pp. 527)
8. **D.** (p. 527)
9. **D.** (pp. 528–529)
10. **A.** (p. 529)
11. **A.** (p. 530)
12. **D.** (p. 531)
13. **A.** (p. 530–532) Note: Three errors in A: not try, but prearrange; not any representative, but the watch supervisor; outside the plant. B: only has one error: should escort them to site. C: has an error also: there is a firefighter life hazard present once we arrive.
14. **C.** (p. 531–532)
15. **B.** (p. 532)
16. **B.** (pp. 532–533)
17. **C.** (pp. 534)
18. **B.** (pp. 535–536) Shut down the smaller individual breakers in the panel first, before opening the main breaker.
19. **B.** (pp. 537–539) The safety zone should be at least one full wire span away from any break.
20. **B.** (p. 540–541) fluorescent, not incandescent

22 — STRUCTURAL COLLAPSE

1. **B.** (p. 543)
2. **A.** (p. 544)
3. **C.** (pp. 544–546)
4. **D.** (pp. 544–545, 488–489)
5. **B.** (pp. 545)
6. **A.** (p. 547) (#2) Steel sags and twists at 1,500°F. (#3) Shrinking can cause collapse by pulling, not pushing.
7. **D.** (p. 547)
8. **C.** (p. 547–548)
9. **D.** (p. 548)
10. **C.** (pp. 550)
11. **B.** (pp. 549–552)
12. **C.** (pp. 553–555)
13. **B.** (p. 553–554)
14. **A.** (pp. 555)
15. **A.** (p. 555) audible, not visual
16. **D.** (p. 556)

17. **D.** (pp. 557)
18. **C.** (pp. 558)
19. **B.** (pp. 558)
20. **C.** (p. 561–562)
21. **C.** (p. 560)
22. **A.** (p. 559)
23. **B.** (p. 560)
24. **B.** (p. 564)
25. **D.** (pp. 564) Note: Problem occupancies involve chemicals.
26. **D.** (p. 565) Note: Account for all victims, not just live persons.
27. **C.** (p. 566–568)
28. **C.** (p. 567)
29. **A.** (p. 567)
30. **C.** (p. 568)
31. **D.** (p. 568)
32. **D.** (pp. 570–572)
33. **B.** (p. 572) Remove all nonessential personnel.

23— FIRE DEPARTMENT ROLES IN TERRORISM AND HOMELAND SECURITY

1. **C.** (p. 582)
2. **A.** (p. 583)
3. **D.** (pp. 580–581)
4. **B.** (p. 580)
5. **C.** (p. 588)
6. **D.** (p. 600)
7. **B.** (pp. 605–606) Only statement 4 is correct.
8. **A.** (p. 610–611)
9. **B.** (pp. 611)
10. **D.** (p. 611)
11. **A.** (pp. 612)
12. **D.** (pp. 613–614)
13. **B.** (p. 615)
14. **A.** (pp. 617–618)
15. **B.** (pp. 618)
16. **D.** (p. 605–606)
17. **A.** (pp. 606)
18 **D.** (p. 607)
19. **D.** (p. 609)
20. **A.** (p. 589–590)

FINAL EXAMINATION

1. **D.** (p. 11)
2. **B.** (p. 16–17)
3. **D.** (pp. 23–24)
4. **A.** (p. 17–18)
5. **A.** (p. 40)
6. **C.** (p. 42)
7. **C.** (p. 47) NO occupants
8. **B.** (pp. 54)
9. **C.** (pp. 76)
10. **C.** (p. 10)
11. **D.** (pp. 29–30)
12. **A.** (p. 49–50) This third-stage fire with its backdraft potential occurs with occupied apartments above. The potentially heavy fire demands heavy caliber stream placement for rapid knockdown. B: If faced with potential backdraft, take the show windows after the engine has bled its line, but while they are off to one side to avoid blast damage. Better yet, let the master stream on the pumper across the street blow the windows in. C: After the show windows are vented, there is no chance for an indirect attack. D: As this is an occupied building, you must vent the upper floors ASAP while conducting searches.
13. **B.** (p. 62)
14. **D.** (pp. 76–354) Fire is in rear, house is *occupied*, no mention of exposure problem. Protection of exits must receive priority to permit search and prevent interior fire spread.
15. **A.** (pp. 369–370)
16. **D.** (pp. 69–75)
17. **B.** (pp.75–78) Note: Haul up to the floor or landing below, then take stairs to the fire floor.
18. **C.** (pp. 100)
19. **B.** (pp. 107–109)
20. **B.** (pp. 33) 100 × 25 ft = 2,500 sq ft involved × 50 gpm per 100 sq ft (lumber storage = heavy fire load) = 1, 250 gpm
21. **D.** (p. 110–111) Handlines of different diameters, lengths, and different nozzle types have very different pressure requirements and could pose serious problems.
22. **D.** (pp. 115,442)
23. **C.** (pp. 127)
24. **A.** (p. 127)
25. **C.** (pp. 133–134) For choice 4, send two members; for choice 5, investigate first.
26. **D.** (pp. 141–155) Note: Choice 4 might depend on the floor of the fire, but it says engine pressure, and 100 psi is too low.
27. **C.** (pp. 121)
28. **D.** (pp. 153–154)
29. **B.** (pp. 125–126)
30. **B.** (pp. 161)

31. **B.** (pp. 172–724)

32. **C.** (pp.)

33. **B.** (p.) Note: Use an aerial ladder or ladder tower (not tower ladders or snorkels) for removing many firefighters from a roof or other area.

34. **A.** (pp. 175) Note: B: 15° to 20°, not 45°; C: backed in; D: 15 ft past the near end, so you can drive forward if needed.

35. **D.** (p. 199)

36. **C.** (pp. 213)

37. **D.** (pp. 215)

38. **B.** (pp. 217)

39. **C.** (pp. 239)

40. **A.** (p. 343)

41. **D.** (pp. 258) Note: A: Fire is in cellar!

42. **D.** (pp. 242)

43. **C.** (p. 237)

44. **A.** (pp. 255–256) Note: D: The original roof will prevent you from pushing the ceiling down.

45. **B.** (pp. 336–337)

46. **B.** (pp. 336–337, 423–424)

47. **D.** (pp. 448–449, 253)

48. **B.** (pp. 257–260) Note: narrowest, not widest

49. **C.** (pp. 248–249, 405–408)

50. **A.** (pp. 250–251) The first is to have an alternate escape route. The wind is second.

51. **C.** (pp. 278–280)

52. **B.** (pp. 309–312)

53. **D.** (p. 276)

54. **C.** (p. 277) A two-person team is required for each apartment or moderate size store or office under light smoke conditions. More are required if smoke is thick, or if heavy stock or cluttered conditions are encountered.

55. **A.** (p. 289) Note: She's lost, not in danger of being overrun by fire. Air conservation is critical.

56. **D.** (p. 287)

57. **C.** (p. 300)

58. **B.** (p. 287) Note: 2: We have to try to go above to search and check for extension. 4: Not always possible or practical, e.g., for a third-floor fire, stretch dry to a safe area.

59. **B.** (p. 287–288)

60. **D.** (pp. 382)

61. **B.** (p. 388)

62. **A.** (pp. 372–373) Note: B: Also requires a line on first floor. C: Poke small examination opening in each bay; don't bother with whole length initially. Don't bother with window and door bays, because they act as fire stops. D: *Requires* roof venting.

63. **C.** (p. 401)

64. **D.** (p. 397)

65. **A.** (p. 392)

66. **C.** (pp. 412–413) Note: All doors

must be positioned the same as they will be on the fire floor; otherwise you do not get a true picture of whether the wind will blow in on the fire.

67. **A.** (pp. 412–415)

68. **B.** (pp. 244–246, 406–407) Note: A: Not yet; be sure no fire in pipe chase or on the way up to the cockloft first, and never descend the interior stair until after the fire is under control. C: Not for ground-floor fire. D: Not unless fire is in the cockloft.

69. **D.** (pp. 427–428)

70. **C.** (pp. 419–420. 426)

71. **B.** (pp. 419–422)

72. **D.** (p. 431)

73. **D.** (pp. 18, 434)

74. **C.** (p. 437)

75. **B.** (pp. 373–376)

76. **D.** (pp. 441–442)

77. **C.** (pp. 434–436)

78. **B.** (p. 431)

79. **B.** (p. 373–374)

80. **B** (pp. 343–345 and 431–432)

81. **C.** (p. 460)

82. **C** (p. 462–463)

83. **C.** (pp. 466–468) Note: A: Firefighter's service elevators are affected the same as all others. B: Avoid freight elevators. D: After precautionary stops every five floors, go to two floors below the fire.

84. **B.** (pp. 461–464)

85. **D.** (pp. 479–482 and 544–547)

86. **B.** (pp. 513)

87. **D.** (pp. 508–509)

88. **C.** (pp. 45–46) After cooling the overhead for 5–10 seconds, the stream should be lowered to cool the rest of the burning material in the room, and sweep the floor of hot debris.

89. **A.** (p. 289) Seek any escape route, do not sit still—that's for when fire is *not* chasing you.

90. **B.** (pp. 258–260) Note: A: 3 ft, C: 3–4 ft, D: *fire side*

91. **B.** (p. 275) Note: Penetrate to the fire and then work back.

92. **A.** (p. 52–53) Note: B: Put the water on the exposure, not in a water curtain. C: Coat the *exposure*, not the fire building. D: Use straight streams from a distance only, when hitting windows for exposure protection.

93. **D.** (pp. 517–518) Note: B: *fluorescent*, not incandescent; C: where first detected

94. **D.** (pp. 444)

95. **C.** (p. 552–555)

96. **C.** (pp. 562–564)

97. **A.** (pp. 569–570)

98. **C.** (pp. 572–573) Note: Use only essential personnel and rotate them frequently.

99. **D.** (p. 608f609)

100. **D.** (pp. 599–600)

101. **B.** (pp. 8)

102. **D.** (p. 36) High-rise would have a standpipe stretch, not a long handline from a hosebed.

103. **B.** (p. 16–19) extent of the fire.

104. **D.** (pp. 27) Class 5 is not fire-resistive construction.

105. **A.** (pp. 33)

106. **C.** (pp. 42–43) The nozzle operator and officer should look.

107. **C.** (pp. 50–51) small hole

108. **D.** (pp. 50–51) Backdrafts can occur in other spaces within the fire building.

109. **B.** (pp. 53)

110. **D.** (pp. 132–134)

111. **A.** (pp. 234–235)

112. **D.** (pp. 238)

113. **A.** (pp. 238)

114. **D.** (pp. 337)

115. **B.** (pp. 286–287)

116. **C.** (pp. 322)

117. **C.** (pp. 347–348)

118. **D.** (p. 348)

119. **C.** (pp. 345–346)

120. **C.** (p. 381)

121. **B.** (p. 403)

122. **D.** (pp. 420)

123. **B.** (p. 427)

124. **D.** (p. 437)

125. **D.** (pp.)

126. **D.** (p. 480)

127. **C.** (p. 392) several (three or more) door bells, mailboxes, etc.

128. **C.** (p. 394) Send someone who has seen this shaft and knows where to look if possible. And be sure to send someone to the top floor as well. Don't forget the base of the shaft, but runoff from hose streams operating in the shaft delay extension downward.

129. **C.** (p. 404)

130. **A.** (p. 408)

131. **D.** (pp. 3411–415)

132. **B.** (p. 380)

133. **A.** (p. 381)

134 **C.** (p. 387)

135. **A.** (p. 306–307)

136. **B.** (pp. 319–320)

137. **D.** (pp. 485–486)

138. **B.** (pp. 462)

139. **C.** (pp. 499–503)

140. **C.** (pp. 535–537)

141. **D.** (pp. 558)

142. **A.** (p. 549)

143. **C.** (pp. 597)

144. **D.** (p. 7–8)

145. **D.** (p. 16)

146. **B.** (pp. 17–18)

147. **A.** (pp. 24–27)

148. **A.** (pp. 29–30)

149. **A.** (pp. 19–23)

150. **A.** (pp. 41)

151. **D.** (p. 55)

152. **D.** (pp. 63)

153. **A.** (pp. 367–368)

154. **C.** (pp. 66)

155. **D.** (p. 67)

156. **B.** (pp. 72)

157. **C.** (pp. 101–114)

158. **D.** (p. 122)

159. **D.** (p. 156)

160. **D.** (p. 165)

161. **C.** (pp. 184–185)

162. **D.** (pp. 192)

163. **D.** (pp. 195–197)

164. **A.** (p. 218)

165. **B.** (pp. 240)

166. **C.** (pp. 241–244)

167. **B.** (pp. 257–261)

168. **C.** (p. 268)

169. **B.** (p. 278) Search where these items will land.

170. **B.** (pp. 280–281)

171. **A.** (p. 289)

172. **B.** (pp. 286–287)

173. **C.** (pp. 295) A belay is only used in training.

174. **C.** (p. 335)

175. **D.** (pp. 337)

176. **D.** (p. 340) Note: B: is true, but the question asked about firefighting operations.

177. **B.** (p. 379) Note: D: Bedrooms are the most likely place to find victims any time of the day.

178. **A.** (pp. 382–383)

179. **C.** (p. 384) Note: Both I-beams and soil-pipe chases go to the cockloft.

180. **C.** (pp. 406–407)

181. **C.** (p. 412)

182. **C.** (p. 419)

183. **A.** (pp. 422)

184. **D.** (pp. 422)

185. **C.** (p. 431)

186. **C.** (p. 432)

187. **D.** (p. 449–451)

188. **D.** (pp. 452–460)

189. **B.** (p. 481)

190. **D.** (pp. 482–485)

191. **D.** (pp. 496,498–500)

192. **D.** (p. 508)

193. **A.** (pp. 524)

194. **B.** (p. 527)

195. **B.** (pp. 532–533)

196. **B.** (pp. 537–539)

197. **B.** (p. 544)

198. **C.** (p. 547–548)

199. **C.** (pp. 580–581)

200. **A.** (pp. 606)

PART V

ANSWER SHEETS

01 GENERAL PRINCIPLES OF FIREFIGHTING

Name ______________________ Date ______________________

Class ______________________ Instructor ______________________

1. ____
2. ____
3. ____
4. ____
5. ____
6. ____
7. ____
8. ____
9. ____
10. ____

02 SIZE-UP

Name ______________________ Date ______________

Class ______________________ Instructor ______________

1. ____
2. ____
3. ____
4. ____
5. ____
6. ____
7. (1) C ______________________
   (2) O ______________________
   (3) A ______________________
   (4) L ______________________
   (5) W ______________________
   (6) A ______________________
   (7) S ______________________
   (8) W ______________________
   (9) E ______________________
   (10) A ______________________
   (11) L ______________________
   (12) T ______________________
   (13) H ______________________

8. _____

9. _____

10. _____

11. _____

12. _____

13. _____

14. _____

15. _____

16. _____

17. _____

18. _____

19. _____

20. _____

21. (1) ____

(2) ____

____

(3) ____

____

22. _____

23. _____

24. _____

25. _____

03 ENGINE COMPANY OPERATIONS

Name ______________________________ Date ______________________

Class ______________________________ Instructor ________________

1. ____

2. ____

3. ____

4. ____

5. ____

6. ____

7. ____

8. ____

9. ____

10. ____

11. ____

12. ____

13. ____

14. ____

15. ____

16. ____

04 HOSELINE SELECTION, STRETCHING, AND PLACEMENT

Name ______________________ Date ______________________

Class ______________________ Instructor ______________________

1. ____
2. ____
3. ____
4. ____
5. ____
6. ____
7. ____
8. ____
9. ____
10. ____
11. ____
12. ____
13. ____
14. ____
15. ____
16. ____
17. ____
18. ____
19. ____
20. ____
21. ____
22. ____

05 WATER SUPPLY

Name ______________________ Date ______________________

Class ______________________ Instructor ______________________

1. _____
2. _____
3. _____
4. _____
5. _____
6. _____
7. _____
8. _____
9. _____
10. _____
11. _____
12. _____
13. _____
14. _____
15. _____
16. _____
17. _____
18. _____
19. _____

06 SPRINKLER SYSTEMS AND STANDPIPE OPERATIONS

Name ______________________ Date ______________________

Class ______________________ Instructor ______________________

1. ____
2. ____
3. ____
4. ____
5. ____
6. ____
7. ____
8. ____
9. ____
10. ____
11. ____
12. ____
13. ____
14. ____
15. ____
16. ____
17. ____
18. ____
19. ____
20. ____
21. ____
22. ____
23. ____
24. ____
25. ____
26. ____
27. ____
28. ____
29. ____
30. ____
31. ____
32. ____
33. ____

07 LADDER COMPANY OPERATIONS

Name ______________________ Date ____________________

Class ______________________ Instructor ________________

1. ____
2. ____
3. ____
4. ____
5. ____
6. ____
7. ____
8. ____
9. ____
10. ____
11. ____
12. ____
13. ____
14. ____
15. ____
16. ____
17. ____
18. ____
19. ____
20. ____
21. ____
22. ____
23. ____
24. ____
25. ____
26. ____
27. ____
28. ____
29. ____
30. ____
31. ____
32. ____
33. ____

08 FORCIBLE ENTRY

Name ______________________ Date ______________________

Class ______________________ Instructor ______________________

1. ____
2. ____
3. ____
4. ____
5. ____
6. ____
7. ____
8. ____
9. ____
10. ____
11. ____
12. ____
13. ____
14. ____
15. ____
16. ____
17. ____
18. ____
19. ____
20. ____
21. ____
22. ____
23. ____
24. ____
25. ____
26. ____
27. ____
28. ____
29. ____
30. ____

09 VENTILATION

Name ______________________ Date ____________________

Class ______________________ Instructor ________________

1. ____
2. ____
3. ____
4. ____
5. ____
6. ____
7. ____
8. ____
9. ____
10. ____
11. ____
12. ____
13. ____
14. ____
15. ____
16. ____
17. ____
18. ____
19. ____
20. ____
21. ____
22. ____
23. ____
24. ____
25. ____
26. ____
27. ____
28. ____
29. ____
30. ____
31. ____
32. ____

33. ____

34. ____

35. ____

36. ____

37. ____

38. ____

10 SEARCH AND RESCUE

Name ______________________ Date ______________________

Class ______________________ Instructor ______________________

1. ____
2. ____
3. ____
4. ____
5. ____
6. ____
7. ____
8. ____
9. ____
10. ____
11. ____
12. ____
13. ____
14. ____
15. ____
16. ____
17. ____
18. ____
19. ____
20. ____

11 FIREFIGHTER SURVIVAL

Name ____________________ Date ____________________

Class ____________________ Instructor ____________________

1. ____
2. ____
3. ____
4. ____
5. ____
6. ____
7. ____
8. ____
9. ____
10. ____
11. ____
12. ____
13. ____
14. ____
15. ____
16. ____
17. ____
18. ____
19. ____
20. ____
21. ____
22. ____
23. ____
24. ____
25. ____
26. ____
27. ____
28. ____
29. ____
30. ____
31. ____
32. ____

33. ____

34. ____

35. ____

36. ____

37. ____

38. ____

39. ____

40. ____

12 OPERATIONS IN LIGHTWEIGHT BUILDINGS

Name ______________________ Date ______________________

Class ______________________ Instructor ______________________

1. ____
2. ____
3. ____
4. ____
5. ____
6. ____
7. ____
8. ____
9. ____
10. ____
11. ____
12. ____
13. ____
14. ____
15. ____
16. ____
17. ____
18. ____
19. ____
20. ____
21. ____
22. ____
23. ____
24. ____
25. ____

13 BELOW-GRADE FIRES: BASEMENTS AND SUBBASEMENTS, CELLARS AND SUBCELLARS, AND CRAWL SPACES

Name ______________________ Date ______________________

Class ______________________ Instructor ______________________

1. ____
2. ____
3. ____
4. ____
5. ____
6. ____
7. ____
8. ____
9. ____
10. ____
11. ____
12. ____
13. ____

14 PRIVATE DWELLINGS

Name ______________________ Date ______________________

Class ______________________ Instructor ______________________

1. _____
2. _____
3. _____
4. _____
5. _____
6. _____
7. _____
8. _____
9. _____
10. _____
11. _____
12. _____
13. _____
14. _____
15. _____
16. _____
17. _____
18. _____

19.

15 MULTIPLE DWELLINGS

Name ______________________ Date ______________________

Class ______________________ Instructor ______________________

1. ____
2. ____
3. ____
4. ____
5. ____
6. ____
7. ____
8. ____
9. ____
10. ____
11. ____
12. ____
13. ____
14. ____
15. ____
16. ____
17. ____
18. ____
19. ____
20. ____
21. ____
22. ____
23. ____
24. ____
25. ____
26. ____
27. ____
28. ____
29. ____
30. ____
31. ____
32. ____
33. ____

16 GARDEN APARTMENT AND TOWNHOUSE FIRES

Name ______________________________ Date _______________________

Class _______________________________ Instructor __________________

1. ____
2. ____
3. ____
4. ____
5. ____
6. ____
7. ____
8. ____
9. ____
10. ____
11. ____
12. ____
13. ____
14. ____
15. ____
16. ____
17. ____
18. ____

17 STORE FIRES—TAXPAYERS AND STRIP MALLS

Name ______________________________ Date _______________________

Class _______________________________ Instructor __________________

1. ____
2. ____
3. ____
4. ____
5. ____
6. ____
7. ____
8. ____
9. ____
10. ____
11. ____
12. ____
13. ____
14. ____
15. ____
16. ____
17. ____
18. ____
19. ____
20. ____
21. ____
22. ____
23. ____
24. ____
25. ____

18 HIGH-RISE OFFICE BUILDINGS

Name ______________________ Date ______________________

Class ______________________ Instructor ______________________

1. ____
2. ____
3. ____
4. ____
5. ____
6. ____
7. ____
8. ____
9. ____
10. ____
11. ____
12. ____
13. ____
14. ____
15. ____
16. ____
17. ____
18. ____
19. ____
20. ____
21. ____
22. ____
23. ____
24. ____
25. ____
26. ____
27. ____
28. ____
29. ____
30. ____
31. ____
32. ____

33. ____

34. ____

35. ____

36. ____

37. ____

38. ____

39. ____

40. ____

19 BUILDINGS UNDER CONSTRUCTION, RENOVATION, AND DEMOLITION

Name ______________________ Date ______________________

Class ______________________ Instructor ______________________

1. ____
2. ____
3. ____
4. ____
5. ____
6. ____
7. ____
8. ____
9. ____
10. ____
11. ____
12. ____
13. ____
14. ____
15. ____
16. ____
17. ____
18. ____
19. ____
20. ____

20 FIRE-RELATED EMERGENCIES: INCINERATORS, OIL BURNERS, AND GAS LEAKS

Name ______________________________ Date ______________________

Class ______________________________ Instructor ____________________

1. ____
2. ____
3. ____
4. ____
5. ____
6. ____
7. ____
8. ____
9. ____
10. ____
11. ____
12. ____
13. ____
14. ____
15. ____
16. ____
17. ____
18. ____
19. ____
20. ____
21. ____
22. ____
23. ____
24. ____
25. ____
26. ____
27. ____
28. ____
29. ____
30. ____
31. ____
32. ____
33. ____

21 ELECTRICAL FIRES AND EMERGENCIES

Name ____________________ Date ____________________

Class ____________________ Instructor ____________________

1. _____
2. _____
3. _____
4. _____
5. _____
6. _____
7. _____
8. _____
9. _____
10. _____
11. _____
12. _____
13. _____
14. _____
15. _____
16. _____
17. _____
18. _____
19. _____
20. _____

22 STRUCTURAL COLLAPSE

Name ______________________ Date ______________________

Class ______________________ Instructor ______________________

1. _____
2. _____
3. _____
4. _____
5. _____
6. _____
7. _____
8. _____
9. _____
10. _____
11. _____
12. _____
13. _____
14. _____
15. _____
16. _____
17. _____
18. _____
19. _____
20. _____
21. _____
22. _____
23. _____
24. _____
25. _____
26. _____
27. _____
28. _____
29. _____
30. _____
31. _____
32. _____
33. _____

23 FIRE DEPARTMENT ROLES IN TERRORISM AND HOMELAND SECURITY

Name ______________________ Date ______________________

Class ______________________ Instructor ______________________

1. ____
2. ____
3. ____
4. ____
5. ____
6. ____
7. ____
8. ____
9. ____
10. ____
11. ____
12. ____
13. ____
14. ____
15. ____
16. ____
17. ____
18. ____
19. ____
20. ____

FINAL EXAMINATION

Name ______________________________ Date ______________________

Class ______________________________ Instructor ______________________

1. ____
2. ____
3. ____
4. ____
5. ____
6. ____
7. ____
8. ____
9. ____
10. ____
11. ____
12. ____
13. ____
14. ____
15. ____
16. ____
17. ____
18. ____
19. ____
20. ____
21. ____
22. ____
23. ____
24. ____
25. ____
26. ____
27. ____
28. ____
29. ____
30. ____
31. ____
32. ____

33. ____
34. ____
35. ____
36. ____
37. ____
38. ____
39. ____
40. ____
41. ____
42. ____
43. ____
44. ____
45. ____
46. ____
47. ____
48. ____
49. ____
50. ____
51. ____
52. ____
53. ____
54. ____
55. ____
56. ____
57. ____
58. ____
59. ____
60. ____
61. ____
62. ____
63. ____
64. ____
65. ____
66. ____
67. ____
68. ____
69. ____
70. ____
71. ____
72. ____
73. ____
74. ____
75. ____
76. ____
77. ____
78. ____
79. ____
80. ____
81. ____
82. ____
83. ____
84. ____
85. ____
86. ____
87. ____
88. ____
89. ____
90. ____
91. ____
92. ____
93. ____
94. ____
95. ____
96. ____
97. ____
98. ____
99. ____
100. ____
101. ____
102. ____
103. ____
104. ____

105. ____
106. ____
107. ____
108. ____
109. ____
110. ____
111. ____
112. ____
113. ____
114. ____
115. ____
116. ____
117. ____
118. ____
119. ____
120. ____
121. ____
122. ____
123. ____
124. ____
125. ____
126. ____
127. ____
128. ____
129. ____
130. ____
131. ____
132. ____
133. ____
134. ____
135. ____
136. ____
137. ____
138. ____
139. ____
140. ____
141. ____
142. ____
143. ____
144. ____
145. ____
146. ____
147. ____
148. ____
149. ____
150. ____
151. ____
152. ____
153. ____
154. ____
155. ____
156. ____
157. ____
158. ____
159. ____
160. ____
161. ____
162. ____
163. ____
164. ____
165. ____
166. ____
167. ____
168. ____
169. ____
170. ____
171. ____
172. ____
173. ____
174. ____
175. ____
176. ____

177. ____

178. ____

179. ____

180. ____

181. ____

182. ____

183. ____

184. ____

185. ____

186. ____

187. ____

188. ____

189. ____

190. ____

191. ____

192. ____

193. ____

194. ____

195. ____

196. ____

197. ____

198. ____

199. ____

200. ____